Sustainable and Smart Farming Eco-systems

EMERALD STUDIES IN POLITICS AND TECHNOLOGY

Series Editor: Anna Visvizi

This series focuses broadly on the intersection of politics and technology. Its objective is to identify and explore the critical junctions where politics and information and communication technology (ICT) intersect to showcase the opportunities, raise awareness, and preempt impending risks for our societies.

The series has a broad scope and addresses a variety of topics, including but not limited to the following: cyber-intelligence; government analytics; user-generated data and its impact on human society; technology in healthcare and public services; quantitative measures in political discourse; public engagement with politics through technology, for example, blogs, social media; freedom of the internet; text mining; e-participation in politics and digital diplomacy; international trade on the ICT market; information security risks; political communication in online social networks; big data; e-government and e-democracy; digital activism; ICT in developing nations; digital media; smart cities; disruptive effects of technology in politics; internet governance; citizen journalism; the politics of migration and ICT; and the European Union and ICT.

Sustainable and Smart Farming Eco-systems: Changing Landscapes in the Age of Technology

EDITED BY

ASHISH GUPTA
South Asian University, India

AND

TAVISHI TEWARY
FORE School of Management, India

emerald
PUBLISHING

United Kingdom – North America – Japan – India – Malaysia – China

Emerald Publishing Limited
Emerald Publishing, Floor 5, Northspring, 21-23 Wellington Street, Leeds LS1 4DL.

First edition 2026

British Library Cataloguing in Publication Data
A catalogue record for this book is available from the British Library

ISBN: 978-1-83608-093-0 (Print)
ISBN: 978-1-83608-092-3 (Online)
ISBN: 978-1-83608-094-7 (Epub)

Printed and bound by CPI Group (UK) Ltd, Croydon, CR0 4YY

Contents

About the Editors

Ashish Gupta is presently working as Senior Associate Professor in the Marketing Area at the Faculty of Management, South Asian University, New Delhi (an international university established by the Government of South Asian Association for Regional Cooperation (SAARC) Nations). He is currently on Lien from Indian Institute of Foreign Trade, New Delhi, an autonomous institution of the Ministry of Commerce and Industry, Government of India. His research interests are in marketing, customer relationship management, service quality, branding, and consumer behaviour. He has over 15+ years of teaching, research, and consulting experience in the area of marketing and management. He has more than 60+ research papers, including case studies to his credit, published in international journals indexed in Australian Business Deans Council (ABDC), Social Sciences Citation Index (SSCI), and Scopus, like the *Journal of Consumer Marketing*; *Journal of Retailing and Consumer Services*; *Asia Pacific Journal of Marketing and Logistics*; *International Journal of Emerging Markets*; *Journal of Internet Commerce*; *Journal of Promotion Management*; *International Journal of Quality and Service Sciences*; *Young Consumers*; *American Business Review*; *Emerald Emerging Markets Cases*; *Journal of Direct, Data and Digital Marketing Practice*, etc.

Tavishi Tewary is currently working as Assistant Professor in Economics. Previously, she served at the Jaipuria Institute of Management, Noida, Uttar Pradesh. She earned her PhD degree in International Economics and Trade from Indian Institute of Management, Mumbai (erstwhile NITIE Mumbai). She has to her credit many national and international publications. She has experience in policy research and trade impact assessment. She has also conducted Faculty Development Programmes (FDPs) on data analysis using SPSS. In academics, she had been associated with teaching at Universidad de las Américas Puebla (UDLAP), Mexico. She has earned her PhD in Economics from Indian Institute of Management, Mumbai (erstwhile NITIE, Mumbai). Her research interests' areas are international economics, sustainability, renewable energy, and macroeconomics modelling.

About the Editors

About the Contributors

Krishnashree Achuthan is a distinguished academic researcher and Dean of Postgraduate Programs at Amrita Vishwa Vidyapeetham as well as CEO of Amrita Technology Business Incubator, renowned for her extensive contributions to the fields of technology, innovation, entrepreneurship, and social reforms. With over 150 publications and 34 patents to her credit, her work is characterised by a unique blend of technical expertise along with an understanding of the societal implications of technology. Her efforts have not only advanced academic knowledge but have also had a tangible impact on policymaking and the broader discourse on technology and social issues. She has served as one of the key coordinators for the National Mission on Internet and Communication Technologies by the Government of India for over a decade. ORCID: https://orcid.org/0000-0003-2618-0882.

Arpan Anand is an Assistant Professor of Marketing at Jaipuria Institute of Management, Noida. He holds an MBA, MCom, and is University Grants Commission – National Eligibility Test (UGC-NET) qualified in Management. With prior experience in the banking sector, he brings practical insights into his teaching and research. His academic work spans marketing, family business, sustainability, and tourism, with publications in reputed journals such as the *International Journal of Emerging Markets* and *Journal of Service Research*. He also leads corporate training programmes for organisations like BHEL, Airtel, and Axis Bank. He was recognised as the Best Faculty at Jaipuria for the academic year 2019–2020.

Janardhana Anjanappa is a distinguished climate finance expert, researcher, and advisor with over 15 years of experience at the intersection of climate change, climate finance, sustainable development, and clean energy. He has implemented high-impact projects supported by global organisations such as the Ministry of Environment–Japan, Green Climate Fund, United Nations Development Programme, United Nations Environment Programme, World Vision, Stockholm Environment Institute (SEI) Asia, and the International Labour Organization (Geneva). His contributions span just transition frameworks, energy finance de-risking, adaptation planning, and technology transfer across Asia, Africa, and Latin America.

Arushi Bathla is a Lecturer at the School of Liberal Arts, O. P. Jindal Global University, Sonipat, Haryana. She holds PhD in General Management and Strategy at the Indian Institute of Foreign Trade (IIFT), New Delhi, and MBA in Public Policy and Social Entrepreneurship from Ambedkar University Delhi and a BA in Economics from the University of Delhi. Her research focuses on design thinking, pedagogical innovation, sustainability, and educational leadership. At IIFT, she combines robust research with hands-on workshops; previously, she led design-thinking initiatives at Ambedkar University and conducted national-level educational training. ORCID: https://orcid.org/0000-0002-0980-4721.

Renju Chandran is an Assistant Professor in the School of Contemporary Knowledge Systems. She holds a doctoral degree in Commerce from Mahatma Gandhi University, Kottayam. Her doctoral research focused on the health insurance business in the public and private sectors. She has presented papers at national as well as international conferences and published research papers in peer-reviewed journals. She started her teaching career at St Albert's College (autonomous), Ernakulam. She was working as the Head of the Department of School of Management Studies, Union Christian College, Aluva, before joining Chinmaya Vishwa Vidyapeeth. ORCID: https://orcid.org/0000-0003-1215-1140.

Chetna Chauhan is an Assistant Professor of Supply Chain and Technology at Universidad de los Andes, Colombia. With a strong academic and research background, she is passionate about leveraging statistical tools and emerging technologies to enhance decision-making and innovation within supply chain operations. Her work bridges theory and practice, focusing on data-driven solutions to real-world challenges in logistics and operations. She is committed to advancing research and teaching in supply chain management, contributing meaningfully to both academic and industry-focused discourse.

Deepak Kumar Chauhan is a Professor of Law at the School of Legal Studies, Central University of Punjab, Bathinda, with over 17 years of teaching and 19 years of research experience. Holding degrees in BSc, LLB, LLM, and LLD, he has made significant contributions to the fields of Environmental Law, Constitutional Law, and Drug Abuse Policy. He has led and completed several research projects funded by University Grants Commission (UGC), Central University of Punjab (CUPB), National Institute of Social Defence (NISD), and the Government of India, including evaluations of the National Food Security Act across North Indian states. He has authored 27 research papers, 9 book chapters, and 1 book. He has supervised multiple PhD and LLM dissertations and delivered over 18 invited lectures. As the Founder Member of the Department of Law, he has actively contributed to curriculum development and institutional building. He is also a recipient of the 'Young Scientist Award' by the Ministry of Petroleum and Natural Gas for his work on freshwater sustainability.

Ashish Mohan Dubey is currently an Assistant Professor at Birla Institute of Technology, Mesra, Ranchi (Off Campus-Patna). His research area includes strategic management and strategic human resource management. He holds a PhD in Management from the School of Management Studies, Motilal Nehru

National Institute of Technology, Prayagraj, India. He has more than 18 years of teaching experience at graduate and postgraduate levels and 8 years of experience in guiding and examining graduate and doctoral theses in India. His interest lies in strategic management and project management. He serves on the editorial board of various journals. He has published several papers in international and national journals and conference proceedings. ORCID: https://orcid.org/0000-0003-4894-1208.

Yamini Gupt is a Professor in the Department of Finance and Business Economics, in the University of Delhi with over 25 years of teaching and research experience. She completed her PhD in economics from Clark University, USA. And her research interests lie in the area of environmental and ecological economics and industrial organization. Her current research is on valuation of ecosystem services; agriculture and climate change; economics of the waste sector and climate change; degraded ecosystems and change in livelihood patterns. Her research has been published widely in several high-quality, peer-reviewed journals.

Piyush Gupta is a Faculty in the area of Operations Management. He has done PhD from Indian Institute of Management, Ranchi, and postgraduation in MTech from Indian Institute of Technology, Dhanbad, in Industrial Engineering and Management. He has more than three years of teaching and research experience. He has worked in Fortune Institute of International Business (FIIB), New Delhi. He has published many high-quality papers in top journals such as *Information Technology and People*, *International Journal of Retail and Distribution Management*, *International Journal of Productivity and Performance Management*, to name a few. His area of research interests falls in service operations, behavioural operations research, process analysis, and improvement. He has been involved in doing interdisciplinary research as well. He is also serving as the Associate Editor in Emerald publishing journal.

Isha Jaswal is an Associate Professor of Economics at Vivekananda Institute of Professional Studies (Guru Gobind Singh Indraprastha University (GGSIPU)), Delhi, with over 12 years of teaching and research experience. She earned her PhD in Economics from Jamia Millia Islamia and specialises in developmental economics, international trade, climate economics, and policy analysis. She serves as a Technical Advisor for Infinite Sum Modeling (USA) and contributes her expertise to the Centre for Regional Trade under India's Ministry of Commerce. Recognised for excellence, she received the Alfred Marshall Education Excellence Award (2022). ORCID: https://orcid.org/0000 0002-3241-5388.

Parv Kasana is a PhD candidate (formerly MPhil) in the Department of Geology at the University of Delhi, specialising in geomorphology and fluvial systems. A project scientist in the Delhi University Geomorphology Lab since October 2019, he has authored seven publications and garnered 38 citations for his contributions to landscape evolution, luminescence dating, and tectonic geomorphology. His expertise spans physical geography, fluvial geomorphology, quaternary geology, tectonic geomorphometry, and landscape evolution. Actively engaged in field research and geomorphometric analysis, he investigates the impacts of flood events and tectonic activity on river channel behaviour and landscape dynamics.

Srishty Kasana earned her PhD from the Department of Finance and Business Economics, University of Delhi. She holds an MA in Economics from Ambedkar University of Delhi and a BA (Hons) in Economics from Miranda House. She has presented her research papers at prestigious institutions, including Indian Institute of Technology Bombay, the Indian Institution of Public Administration, the University of Delhi, TERI University, Mahatma Gandhi Central University, and the National Institution of Securities Market, an educational institution established by SEBI. She has published her work in various journals and book chapters and has been invited to deliver lectures on climate change and agriculture.

R. Kavitha is a Professor and Head of the Department of Farm Machinery and Power Engineering at the Agricultural Engineering College and Research Institute, Tamil Nadu Agricultural University, Coimbatore. With over a decade of experience in teaching and research, she leads initiatives in farm machinery, agro-energy, and power systems. She holds a doctorate in Agricultural Engineering and oversees both academic programmes and research projects within her department. She also manages a team of faculty focused on mechanisation innovation to boost agricultural productivity.

Anukaran Khanna joined the School of Computing as an Assistant Professor in November 2024. He completed his Bachelor of Technology in Applied Electronics and Instrumentation Engineering from Uttar Pradesh Technical University, India, in 2007. After BTech, he worked with Golden Cross Pharma Pvt. Ltd (a unit of Cipla Ltd) as an instrumentation engineer in Rorathang, Sikkim. Thereafter, he earned his master's degree specialising in Control and Instrumentation Engineering from the Electrical Engineering Department of Motilal Nehru National Institute of Technology, Allahabad, India, in the year 2013. Before joining the PhD programme at the Department of Electrical and Electronics Engineering, Birla Institute of Technology and Science (BITS) Pilani, India, he was a faculty member in United College of Engineering and Research, Allahabad, India. His research interests lie in nonlinear, robust, and adaptive control techniques as applied to aircraft flight control. ORCID: https://orcid.org/0000-0002-1252-2557.

Prateek Khanna is an Alumnus of Indian Institute of Management Indore and Summer Faculty Research Fellow Programme (SFRF) from Indian Institute of Technology Delhi. He has more than 10 years of teaching and research experience in management education. He contributed various articles and published research papers in national and international journals (ABDC, Emerging Sources Citation Index (ESCI), and Scopus). He has presented various research papers, national and international conferences, and book chapters (Springer, Taylor & Francis, and World Scientific). His areas of interest are consumer behaviour, shopping patterns, sensory branding, and digital consumption. He is a subject expert in Marketing Management for All India Council for Technical Education – Student Learning Assessment Project (AICTE-SLAP) and has completed certification from National Programme on Technology Enhanced Learning (NPTEL). He is a reviewer for various reputed journals like *Marketing Intelligence and Planning* and *Vision: The Journal of Business Perspective*. ORCID: https://orcid.org/0000-0003-2708-0568.

Varsha Khattri is an Associate Professor in the area of Marketing at FORE School of Management, New Delhi, India. With over two decades of academic experience, she has been actively engaged in teaching and research in the areas of brand management, advertising management, consumer behaviour, and crisis communication. Her research papers have been published in several reputed national and international journals. She has been associated with training programmes for organisations such as Punjab National Bank (PNB), the Pacific Asia Travel Association (PATA) India Chapter, Tata Motors, Commonwealth Games, Collabera Services Pvt. Ltd., and Greenlam Industries, among others. She has also taught courses at IMT CDL Ghaziabad and Amity Business School, Noida, India, and has been associated with Management Development Programmes (MDPs) for officers of the Directorate General Resettlement, Ministry of Defence, India, at IIFT Delhi. ORCID: https://orcid.org/0000-0002-9606-1056.

Kokila Lankathilake Konasinghe is Professor of Law and Head of the Department of Public and International Law at the University of Colombo, Sri Lanka. She is the founding Director of the Centre for Environmental Law and Policy and a nationally recognised expert in climate and environmental law. A member of the Commonwealth Futures Climate Research Cohort, she represented Sri Lanka at the United Nations Climate Change COP27 in Egypt. She has led regional research projects under SAARC and the International Red Cross, and currently chairs the drafting committees for Sri Lanka's new National Environmental Act, Climate Change Act, and the National Policy on Living Entities. She has held visiting positions at Erasmus University Rotterdam, University of Granada, and Jindal Global University. She collaborates with Jawaharlal Nehru University and serves as Advisor to the World Youth for Climate Justice. Her scholarly work appears widely in national and international law journals and edited volumes.

Shivani Mehta is an Associate Professor of Economics at Manav Rachna International Institute of Research and Studies, where she contributes to both teaching and research. She leads significant initiatives – including National Agricultural Science Fund (NASF)-funded projects – and actively advocates for economic empowerment through her fieldwork, such as efforts supporting tribal women in Northeast India, recently published in *Frontiers*. A dynamic educator, she engages in faculty development and curriculum advancement, fostering student success and academic excellence. Her interdisciplinary approach integrates hands-on learning with data-driven insights, shaping tomorrow's economists.

Akhilesh Kumar Mishra is an Assistant Professor in the School of Engineering at Institute for Integrated Learning in Management (IILM), Greater Noida. Dr Mishra received his PhD from BITS Pilani, Pilani Campus with a specialisation in Power systems in July 2023. He obtained his BTech (EEE) from Uttar Pradesh Technical University (UPTU) Lucknow and MTech (Control and Instrumentation) from Motilal Nehru National Institute of Technology (MNNIT) Allahabad in 2008 and 2013, respectively. He has more than 10 years of teaching experience at institutions affiliated with Dr. A. P. J. Abdul Kalam Technical University (AKTU) Lucknow and Rajiv Gandhi Proudyogiki Vishwavidyalaya (RGPV) Bhopal. He has authored many research articles and book chapters which have been published in Science Citation Index (SCI), Scopus-indexed journals, and IEEE and International Federation of Automatic Control flagship conferences. He is also an active reviewer of *IEEE Access*, *ISA Transactions*, *IEEE Transactions on Power Systems*, *Electrical Engineering*, *Energy Reports*, and *Arabian Journal of Science and Engineering*. His research interests include load frequency control, renewable energy integration to the grid, intelligent control, fractional order controllers, and forecasting utilising deep learning approaches.

A. P. Mohankumar is an Assistant Professor in the Department of Farm Machinery and Power Engineering at the Agricultural Engineering College and Research Institute, Tamil Nadu Agricultural University (TNAU), Coimbatore. His research focuses on computer-aided design, fluid power systems, and precision farm machinery, including drone-assisted spraying technologies. With a PhD in Farm Machinery from TNAU, he has led several research projects, including the development of a mini-tractor-operated sugarcane detrasher. He has published widely in his field and is an active member of professional bodies like the Indian Society of Agricultural Engineers and the Institution of Engineers. ORCID: https://orcid.org/0000-0003-0840-204X.

Eutychus Ngotho Gichuru, based at College of Education and External Studies, of Makerere University, specializes in Adaptive Management, Higher Education Interdisciplinary Studies and Development, Third Mission, Promotion and Tenure of Faculty in Flagship Universities, Kafkaesque Bureaucracy by Frank Kafka, Fallibilism by Charles Sanders Peirce, Plato Dialectic, Dialectical Reasoning by Hegel, and Herbert Marcuse Philosophy of Repressive tolerance. He has been published in Emerald, Taylor & Francis, and Elsevier. He is a regular contributor to Conversation Africa, and University World News. ORCID: https://orcid.org/0000-0002-5363-5443.

Naresh Prajapati is an Assistant Professor of Law at the School of Law, Bennett University, Greater Noida, India. He has recently submitted his doctoral dissertation titled 'Judicial Approach for Human Rights Protection Amid Climate Change: A Comparative Analysis Across India, Bangladesh, Sri Lanka, and Maldives' at the Central University of Punjab. His research spans international climate change law, human rights, and interdisciplinary legal methodologies. He has published over 20 peer-reviewed articles and book chapters in reputed

Scopus- and ABDC-indexed journals. He has actively presented more than 30 research papers at international and national conferences across South Asia, the Middle East, and Europe. He also serves as Lead Editor for a special issue of the *Journal of International Women's Studies* (Scopus Q2), Senior Editor at the *Africa University Law Review*, and as a reviewer for various academic journals and conferences. His academic work contributes significantly to the intersection of law, environment, and social justice.

Amit Kumar Prasad is a Postdoctoral Associate in the Department of Mechanical Engineering and Materials Science, University of Pittsburgh, Pennsylvania, USA. His current research topic utilises the Angstrom scale and conventional techniques to investigate the relationship between surface roughness and surface properties. He completed his PhD in the Materials Science and Engineering Department at Indian Institute of Technology (IIT) Delhi. His PhD work is on developing novel Mg-RE alloys for automobile applications. He completed his master's degree from IIT Banaras Hindu University (BHU) Varanasi in the Materials Science Department. Earlier, he worked as an Assistant Professor for almost six years at the United College of Engineering and Research (AKTU). ORCID: https://orcid.org/0000-0002-6984-9112.

Rathinavel S. is Doctorate from Tamil Nadu Agricultural University, Coimbatore, India, in the Department of Farm Machinery and Power Engineering. His research is towards sustainable agriculture through variable rate fertiliser application technology assisted by drones. He is also interested in robotics, coconut mechanisation, bioenergy, and extension studies. ORCID: https://orcid.org/0000-0002-6493-6495.

Palak Rawat is a Research Assistant at Manav Rachna International Institute of Research and Studies (Faridabad) since January 2024. She holds a PhD from Manav Rachna International Institute of Research and Studies (MRIIRS) and an MA (Hons) in Economics from Delhi University. She is skilled in quantitative research techniques, including STATA, SPSS, bibliometric analysis, and project management. She combines strong academic credentials with practical research experience, contributing to evidence-based studies in education and social sciences.

Padma Rao Sahib is an Associate Professor in the Faculty of Economics and Business at the University of Groningen's Department of Global Economics & Management. She holds a PhD (1992) and MA (1991) in Economics from Cornell University. Her research focuses on pre-Merger & Acquisition (M&A) processes, organisational learning, and international business, with notable contributions to the study of merger announcement timing, divestiture learning, and acquisition dynamics across sectors like energy, newspapers, and agriculture. She has authored or co-authored over 30 peer-reviewed articles, book chapters, and case studies.

Shridhar M. Samant is a distinguished professional serving as a Mentor in the "Systems Thinking & Environment" stream of the Rishihood University Foundation Programme. With a PhD in Environmental Management, he brings over a decade of rich research and field experience. Throughout his career, he has tackled topics including climate change science and policy, disaster risk reduction (especially frameworks like Sendai and British Broadcasting Corporation (BBC)), sustainable natural-resource governance, food security, smart cities, Environmental, Social, and Governance (ESG), and corporate governance. He is committed to experiential learning and often integrates case studies and fieldwork into his teaching.

Sanitha A. C. is an Assistant Professor in the School of Business and Economics Studies at Chinmaya Vishwa Vidyapeeth (Deemed to be University), Ernakulam. A highly accomplished academic, she cleared the University Grants Commission-National Eligibility Test (UGC-NET) with Junior Research Fellowship (JRF) in Commerce on her very first attempt and holds a Ph.D. from Mahatma Gandhi University, Kottayam, alongside an M.Com in Finance from the University of Calicut and an MBA in Finance from the Central University of Puducherry. Sanitha brings over eight years of rich teaching experience and is a certified Chartered Wealth Manager (CWM) through the American Academy of Financial Management, backed by professional credentials including Cost and Management Accounting (CMA) (Intermediate) from Institute of Cost Accountants of India, a Diploma in Indian and Foreign Accounting (DIFA). As an active research supervisor specializing in behavioral finance, her broader research and teaching interests span taxation, wealth and investment management, Cost and Management Accounting, and Indian Knowledge Systems. Her prolific scholarly contribution includes 25 research papers published in prestigious national and international journals, alongside 31 conference presentations globally. Committed to continuous professional excellence, she has participated in numerous Faculty Development Programs and completed 10 NPTEL courses from premier IITs and IIMs, earning elite distinctions such as NPTEL Believer, Discipline Star, and Enthusiast. Sanitha is also highly proficient in advanced quantitative data analysis using SPSS, AMOS, and SmartPLS.

Sarath Chandran M. C. is an Assistant Professor in the Department of Commerce and Management at Amrita School of Arts, Humanities, and Commerce, Amrita Vishwa Vidyapeetham (Deemed to be University), Amritapuri, India. He holds doctoral degrees in Management and Commerce and is UGC-NET qualified in both disciplines. His research interests lie in banking and finance, green finance, sustainability, and sustainable livelihoods. He has published extensively in reputed national and international journals, including *Discover Sustainability*, *Environmental Development and Sustainability*, and the *International Journal of Agricultural Sustainability*. He is also actively engaged in e-learning content development and has contributed as a subject expert for Quadrant 1 Massive Open Online Courses under the Amrita Ahead Online Program. He has received multiple academic awards, holds granted Indian patents, and regularly presents his research at national and international conferences. He is proficient in quantitative research methods using SPSS, AMOS, and R programming.

Reetika Sehgal is working as Assistant Professor (Human resources/Organizational Behaviour (HR/OB)) in the Department of Management Studies, Dehradun Institute of Technology (DIT) University, Dehradun, since January 2025. She holds a PhD in Management from Birla Institute of Technology, Mesra, Ranchi. Her research area includes human resource management, organisational behaviour, consumer behaviour, general management, and public administration. She has over 12 years of teaching experience at graduate and postgraduate levels. She has published several research papers in international and national journals. She has presented her research work at national and international conferences such as Indian Institutes of Managements, Indian Institutes of Technology, National Institute of Technology (NITs), etc. She has gained various certification courses from the National Program on Technology Enhanced Learning. She has been a reviewer for various national and international journals. She has been honoured with the award 'Academic Excellence' and 'Nation Builder Award' for exemplary contribution towards the betterment of educational objectives. ORCID: https://orcid.org/0000-0002-3659-4766.

Vartul Shrivastava is a doctoral researcher in Information Systems at the Indian Institute of Management Bangalore (IIMB). His research interests span Information and Communication Technology for Development (ICT4D), Artificial Intelligence (AI) anthropomorphism, emotional AI, and companion AI. He has contributed to projects involving large language models, natural language processing, and fuzzy cognitive maps, with work published in journals including the Journal of Environmental Management and SoftwareX. He holds a Bachelor of Technology (BTech) in Computer Science and Engineering from the Indian Institute of Information Technology (IIIT) Vadodara. ORCID: https://orcid.org/0009-0006-4088-6901.

Harsh Shukla is a student of Bachelor of Science – Master of Science (BS-MS) at the Indian Institute of Science Education and Research, Bhopal, India. His research interests include blockchain technologies, natural language processing, machine learning, and deep learning. His work focuses on leveraging federated learning for ensuring privacy and also on neuro-symbolic reasoning approaches to enhance the reasoning capabilities of large language models.

Shekhar Shukla is an Assistant Professor in the Information Systems area at the Indian Institute of Management (IIM) Indore, holding an Fellow Programme in Management (PhD equivalent) from IIM Lucknow and a BE in Computer Science and Engineering from Visvesvaraya Technological University (VTU) Belgaum. Before joining IIM Indore, he taught at S.P. Jain Institute of Management and Research, Mumbai. His research delves into emerging technologies – especially blockchain and artificial intelligence applications in business, governance, and society – and has earned accolades like the Best Paper Award at Shailesh J. Mehta School of Management (SJMSOM), Indian Institute of Technology Bombay, and a Tata Scholarship. His work features in top journals such as *IEEE Transactions on Engineering Management* and *Journal of Research in Interactive Marketing*, covering topics like blockchain for education, healthcare, and e-commerce strategy.

Siva M. is Doctorate from Tamil Nadu Agricultural University, Coimbatore, India, in the Department of Plant Pathology. His research focuses on sustainable management of Plant diseases, particularly Sugarcane Pokkah Boeng disease, through epidemiology, host–pathogen interactions, molecular diagnostics, and endophytic bioformulations. His interests include molecular Plant Pathology, metabolomics, bioinformatics and biological disease management. Published 9 research papers, 4 review papersin peer-reviewed national and international journals, along with 6 book chapters, 4 popular articles and obtained one design patent. Attended several international and national conferences, workshops and trainings. ORCID: https://orcid.org/0000-0003-1457-3275.

A. Surendrakumar is a Professor in the Department of Farm Machinery and Power Engineering at Tamil Nadu Agricultural University (TNAU), Coimbatore, where he has served since 1995. He earned his PhD from TNAU in 2010 and specialises in the design and development of farm machinery. With a portfolio of 16 journal and 6 conference publications, he has contributed to research in sensor-based planters and battery-powered sprayers, guiding doctoral theses such as the sensor-based check-row maize planter (2022) and a battery-powered knapsack air-assisted sprayer (2019). He has led and co-led numerous Indian Council of Agricultural Research (ICAR)- and TNAU-funded projects, including ergonomic tool distribution to Scheduled Class (SC) farmers (2016–2022), vegetable transplanter ergonomics (2016–2019), and vibration studies in mini-tractors (2016–2017). He is a life member of Indian Society of Agricultural Engineers (ISAE) since 2015 and serves on editorial and technical committees, including Uzhavarin Valarum Velanmai and Asian and Pacific Network for Testing of Agricultural Machinery (ANTAM). ORCID: https://orcid.org/0000-0001-5610-5184.

Amarnath Tripathi is currently an Associate Professor at the Jaipuria Institute of Management Noida, India. Previously, he has been a member of the faculty in the School of Business Studies, Sharda University, Greater Noida, Symbiosis School of Economics, Symbiosis International (Deemed University), Pune, and the Institute of Economic Growth, New Delhi. He has also been a Visiting Doctoral Fellow of the Indira Gandhi Institute of Development Research, Mumbai. He completed his PhD in Economics from Banaras Hindu University, Varanasi, and his postdoctorate from the Institute of Economic Growth, New Delhi, under Think Tank, an initiative of the International Development Research Centre, Ottawa, Canada. He has over 13 years of experience in academic and applied economic research. He has published widely, and several of his research articles have appeared in good impact journals like *Food and Nutrition Bulletin*, *Environment Management*, *Climatic Risk Management*, *Journal of Environmental Management* *and Planning*, and *Economic and Political Weekly*. He also writes columns in both print and digital media. ORCID: https://orcid.org/0000-0001-7852-3408.

Kanhaiya Tripathi is the Chair Professor at the Dr B.R. Ambedkar Centre of Excellence, Central University of Punjab, Bathinda. He previously served at Rashtrapati Bhavan as Officer on Special Duty and Editor to the President of India – first under President Pranab Mukherjee and earlier under President Pratibha Patil – editing official publications such as *Bharat Jagao!* and *Rise India!* A prolific author, he has published over 40 books, 150 research articles, and 350 editorial pieces and has held key roles including founder of the Kanhaiya Tripathi's Initiative and advocate for global non-violence. His work spans human rights, social justice, governance, indigenous rights, and environmental policy. A recipient of the National Human Rights Commission (NHRC) creative writing award and a nominee for the Gandhi Peace Foundation's Nobel Peace Award (2020), he actively contributes to academia and public discourse through lectures, media columns, and policy advocacy.

Andrea K. Veira is presently a Lecturer in Science Education at The University of the West Indies (UWI), Cave Hill Campus, Barbados. Her previous career posts include Lecturer in Teacher Education at The UWI Five Islands Campus, Antigua & Barbuda and Crop Scientist and Head of the Caribbean Agricultural Research and Development Institute's (CARDI) Saint Lucia Office/Research Station. She has over 20 years of teaching experience at both secondary and tertiary levels. Her PhD in Biology from The UWI, focussed on agroecology in growing sweet potatoes utilising different plant species in the Caribbean. Her research interests focus on educating to solve food insecurity, sustainable agriculture, integrating information and communication technology in education, student engagement and reflective practice. ORCID: https://orcid.org/0000-0002-3320-8743.

Preface

Agriculture stands at a critical crossroads. With the pressures of climate variability, resource scarcity, population increments, and global market dynamics intensify, the need for sustainable agricultural solutions has become important. Across the world, farmers, researchers, and policymakers are struggling to make agricultural systems more intelligent, resilient, and future-ready to address the challenges of the 21st century. This edited volume brings together a diverse collection of insights, innovations, and interdisciplinary perspectives that collectively illuminate the vast potential of smart and sustainable agriculture.

The chapters in this book explore the evolving ecosystem of agri-technology – spanning internet of things-enabled smart farming, climate-smart practices, robotics and mechatronics, blockchain-enabled supply chains, and artificial intelligence-driven marketing systems. By bridging theory and practice, these contributions elaborate how digital tools, climate-aware policies, and sustainable management strategies can address farmer's challenges, strengthen food systems, and support rural livelihoods.

This book is designed to serve researchers, practitioners, policymakers, students, and all those seeking to understand the transformations of modern agriculture. Each chapter contributes to a broader, collective vision: an agricultural system that is smarter and greener. It is our hope that this volume not only advances scholarly discourse but also inspires meaningful action – encouraging collaborations across disciplines and sectors as we work towards building resilient agricultural futures for generations to come.

Chapter 1

Internet-of-Things (IoT)-powered Smart Farming: Enabling Fields to Communicate

Anukaran Khanna[a], Akhilesh Kumar Mishra[b], Reetika Sehgal[a], Prateek Khanna[c], Amit Kumar Prasad[d] and Ashish Mohan Dubey[e]

[a]*DIT University, Dehradun, Uttarakhand, India*
[b]*IILM University, Greater Noida, Uttar Pradesh, India*
[c]*ICFAI Business School, ICFAI University, Jaipur, Rajasthan, India*
[d]*University of Pittsburgh, Pittsburgh, PA, USA*
[e]*Birla Institute of Technology, Mesra, Ranchi (Patna Campus), Patna, Bihar, India*

Abstract

The emergence of Internet-of-Things (IoT) technologies has brought to the age of "smart farming" in which modern agriculture is more precise, smarter, and guided by data. The objective of this chapter is to examine the impact of the IoT on agriculture. This chapter summarizes numerous types of sensors based on the specific application such as irrigation, pest detection, and crop and soil sensors. The purpose is not only to demonstrate what all these technologies do but also how they benefit farmers at each stage of the crop-growing process; from planting to harvesting and packing it for shipment. It also discusses how unmanned aerial vehicles (UAVs) can handle crop surveillance and yield adjustments for real-time supervision, ensuring proper actions to be taken immediately. It reviews recent IoT architectures and platforms that are used in agriculture. Such platforms are also helping to establish unified information exchange and operation across distinct IoT devices, leading toward an integrated agricultural management system. Utilizing data from a local survey among farmers, this chapter demonstrates how these innovations affect precision agriculture in practicality. As per the findings, it may be concluded that IoT enhances productivity and efficiency through better deployment of resources leading to higher crop yield. The installation of sensor devices

Sustainable and Smart Farming Eco-systems:
Changing Landscapes in the Age of Technology, 1–19

doi:10.1108/978-1-83608-092-320261001

and real-time data collection enables farmers to make appropriate decisions which will help in increasing crop yields while managing resources as efficiently as possible.

Keywords: Agriculture; IoT; smart farming; UAV; crop yield

1. Introduction

Agriculture plays a substantial role in India's economic composition and cultural heritage. It employs between 60% and 70% of the labor force and contributes about 17%–18% of gross domestic product (Gulati & Juneja, 2022). A variety of crops, including wheat, pulses, rice, oilseeds, vegetables, and fruits, are produced by the country's diverse agricultural sector. However, agricultural output regularly drops in its capacity despite the vast base of resources. Productivity is hugely affected by various issues such as farmers' ignorance, poor adoption of technology, and pest and disease outbreaks (Bhargava et al., 2014). The problems get worse by the semi-arid climate, which results in major crop losses (Shinde & Kulkarni, 2017). It is estimated that the total population of the world will increase by approximately 25% and touch the 9 billion marks by 2050 (Tripathi et al., 2019). To meet the demand of such a large population production needs to be doubled. This can only be possible by adopting cutting-edge technologies.

In conventional farming, the farmer needs to continuously visit the crop site to monitor and assess the health of crops as they grow. The health monitoring of crops consumes about 70% of the time which can be easily utilized for fieldwork, the emergence of smart agriculture is paving the way to these obsolete methods' inadequacies (Navulur & Prasad, 2019). The huge scope of agriculture calls for the development of precise and technological solutions that prioritize sustainability and reduce environmental impact. The recent developments in the field of sensing and communication technologies have made remote monitoring of the field possible, which makes the life of farmers a bit easier by keeping an eye on conditions without in person being there. Today, wireless sensor technology provides continuous crop monitoring with high accuracy that helps in identifying the problems at an early stage. From planting and harvesting to storing and transporting, this shift to smart agriculture integrates a variety of cutting-edge instruments (Buyya & Dastjerdi, 2016; Jawahar et al., 2014). Apart from improving operational efficiency, on time data reporting from various sensors cuts down expenses by enabling accurate monitoring. More and more sophisticated drones, harvesters, robotic weed eaters, autonomous tractors, and satellites are assisting modern agricultural practices. In addition to aiding fast data collection and online analysis, these technologies improve conventional equipment. By providing in-depth, precise data for every field, sensor technology supports site-specific agriculture and promotes more effective and sustainable farming methods.

The footprints of Internet-of-Things (IoT) can be found in various sectors like health care, smart city, energy sectors, and residential areas and is now beginning to change the agriculture sector as well (Ayaz et al., 2018; Elijah et al., 2018;

Hi et al., 2019; Lin et al., 2017; Sisinni et al., 2018). IoT applications are still at an early stage, but it is evident that the technology has the potential to improve performance and save inefficiencies. In agriculture, IoT is set to play a significant role due to its extensive abilities. These include but are not limited to the fundamental communication infrastructure which links smart devices – ranging from sensors and vehicles to mobile devices – over the internet. Additionally, IoT also offers a range of services like decision-making, user interface, local and remote data collection, cloud-based intelligent analysis, and automation of agricultural methods. These advancements have the potential to revolutionize agriculture, one of the most inefficient sectors in our economic value chain. By harnessing IoT technologies, India could significantly improve resource management, increase agricultural productivity, and address the gap between growing food demand and supply.

2. Major Applications

Integrating cutting-edge sensing and IoT technologies into agriculture has the potential to revolutionize traditional farming methods. The seamless application of wireless sensors and IoT in smart agriculture can elevate farming practices to new heights previously thought unattainable. By adopting smart agriculture practices as shown in Fig. 1.1, IoT can address many long-standing challenges such as drought management, yield optimization, land suitability, irrigation, and pest control (Bendre et al., 2015).

In the next subsection, we explore how these advanced technologies are improving efficiency at various stages of farming.

2.1. Soil Sampling and Mapping

The crop and soil receive only the precise amount of necessary inputs – such as water, nutrients, fertilizers, and herbicides. To achieve this, soil testing is essential. Conducting a soil test serves several key purposes: it optimizes crop yield, prevents environmental contamination from runoff and excess fertilizers, helps

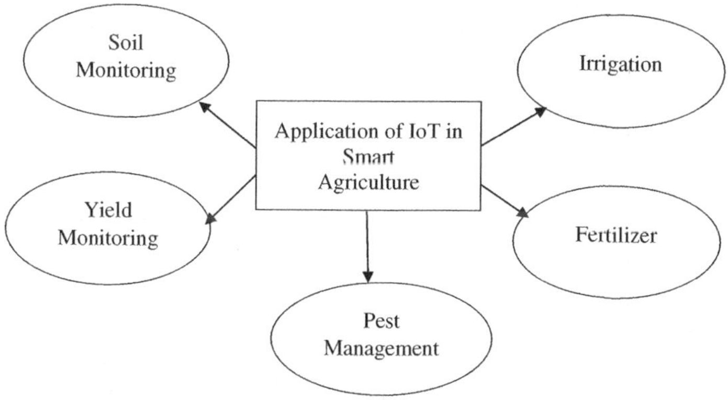

Fig. 1.1. Applications of IoT in Smart Agriculture. *Source*: Authors' compilation.

diagnose plant issues, improves soil nutrient balance, and allows for cost and energy savings by ensuring that only the required amount of fertilizer is applied (Kumar et al., 2021). Physical properties of soil, such as structure, texture, and color, are relatively easy to assess because they can be observed directly (Han et al., 2016). However, the chemical composition of the soil, including factors like pH and nutrient content, is not visible to the naked eye. Since soil fertility is determined by the interplay of biological, chemical, and physical properties – and fertile soil is crucial for healthy crop growth – accurate soil diagnosis through sampling becomes essential.

Farmers can avoid land degradation caused by both under and over fertilization which is difficult to reverse in a short span of time by knowing the nutrient content of the soil. With approximately 12 million hectares of land being degraded per annum, soil depletion and land degradation is one of the major global threats that directly impacted the lives and health of about 1.5 billion people. The primary reasons for this degradation are unsustainable soil management practices that lead to soil erosion, nutrient imbalances, and the loss of soil organic matter (Abdelrahman, 2023). Making prompt decisions, evaluating soil samples, and identifying soil requirements have always been critical tasks for farmers.

Grid mapping is the process of dividing a field or farm into smaller, uniform sections with the use of global positioning system (GPS) technology. As per the desired level of detail, each section of the field covers a particular area, either one acre or one hectare. With the help of advanced tools such as drones fitted with sensors and GPS, the farmers can gather precise data easily. With the help of the data received, they can tailor their farming methods to meet the requirements of each section. Grip mapping approach increases crop production, minimizes environmental impact, and reduces waste by optimizing resource use including fertilizers, water, and pesticides (Mohamed et al., 2021). This technique plays an important role in the development of precision agriculture by not only promoting the sustainable farming practices but also providing farmers data-driven insights which help them make better decisions.

In smart farming, zone mapping is the technique in which a field is divided into discrete zones depending upon the common characteristics like topography, soil properties, and crop performance. Unlike conventional grid mapping, which uses uniform sections, zone mapping aims on detecting areas with similar conditions within the field. This makes it possible to execute more specific management practices. By using advanced technologies such as drones, satellite imagery, and soil sensors, farmers can collect data to measure field variability (Kumar et al., 2022). The data gathered is then studied to define various zones with similar nutrient requirements, soil texture, and moisture levels. Farmers can improve crop yields and manage resources more effectively by varying inputs like fertilizer, pesticides, and water to the exclusive requirements of each zone. This method is a crucial component of precision agriculture.

2.2. *Irrigation*

A major portion, approximately 97%, of water on earth is salty and is found in seas and oceans, this means only 3% of the water on earth is freshwater (Universe

Today, 2014). Out of this freshwater about two-thirds is trapped in polar ice caps and glaciers. The remaining unfrozen freshwater is primarily underground, with only about 0.5% available above ground or in the atmosphere (Musie & Gonfa, 2023). This small portion of freshwater is critical for maintaining ecosystems and fulfilling human needs, as rivers, lakes, and other reservoirs must contain enough to support both. Especially, the agriculture sector utilizes around 70% of this accessible freshwater, with some nations like Brazil consuming up to 75% and some other developing regions using up to 80% (Hoogeveen et al., 2018; Oliveira et al., 2017). Taking the global water crisis and continuous demands from various sectors into account, it is important to make certain that water is distributed effectively. To solve this issue, the focus on conserving water through advanced irrigation techniques is continuously increasing.

The introduction of new technology is expected to transform the future of irrigation. The use of sensors to monitor various parameters such as moisture content of soil, health of crops, and weather conditions helps in better water management. The data received from various sensors are analyzed with the help of a machine learning algorithm, and based on the obtained result, irrigation plans are modified. With the help of IoT-based devices, wastage of water and operating cost can be significantly reduced by adjusting the irrigation schedules (Cruz-Blanco et al., 2014).

2.3. Fertilizer

Fertilization is one of the most important agricultural processes for the development of crop. With the help of technology like real-time sensors, GPS and Geographic Information System (GIS), farmers can restrict the use of fertilizers and reduce wastage and its impact on the environment. Drones and satellites provide useful information on soil and crop conditions, help to make site-specific decisions which further improve the overall efficiency (Farooq et al., 2020; Liu et al., 2018). Other than increasing the crop yields, smart farming supports sustainable agricultural practices.

2.4. Crop Disease and Pest Management

According to assessments from the Food and Agriculture Organization (2019), there are 20%–40% of the crop yield losses annually because of pests and diseases. To overcome this substantial loss, over the past century, pesticides and other agrochemicals have been widely used in agriculture practice. Annually, about two million tons of pesticides are used globally. Overuse of these chemicals is dangerous for the life of both humans and animals, causing environmental damage and contaminating ecosystems (Carvalho, 2017; Waskom et al., 2017). Crop disease and pest management are revolutionized by the introduction of advanced technologies. The detection of disease and pest infestation at an early stage helps farmers to make better decisions, such as applying pesticides only where necessary, which also helps in reducing the use of chemicals. Proper management is made possible by the analysis of the data collected by these sensors using machine learning tools to predict crop health, weather patterns, and soil conditions for any possible outbreaks (Oberti et al., 2016). Smart agriculture improves the effectiveness of disease and pest management and promotes sustainable farming practices.

2.5. Yield Monitoring

Farmers can increase their productivity in terms of both quantity and quality if they switch to modern technology. The use of technology helps to track and analyze the performance of the field in real time and accordingly farmers make decisions. GPS-enabled harvesters, sensors, data analytics altogether a yield monitoring system can precisely assess the crop condition from a section of a field, offering comprehensive information on productivity (Luigi et al., 2015). Farmers can use these data and identify areas that are not performing well and accordingly allocate resources, such as modifying the application of fertilizer or irrigation requirements. Feedback in real time reduces waste, boosts productivity, and eventually results in higher profitability.

3. Major Equipment and Technologies

In contrast to conventional farming, modern agriculture practices are largely dependent on advanced machinery and technologies, such as tractors, harvesters, and robots. Vehicles equipped with GPS and GIS systems perform various tasks like sowing, fertilizing, irrigation, and harvesting allowing them to operate precisely and site-specifically. The integration of these cutting-edge technologies is necessary to the concept of site-specific crop management. Precision agriculture's success depends on the accuracy of data collection, achieved through two main methods. The first method involves using multifunctional imaging devices with remote sensing capabilities, such as satellites, agricultural aircraft, balloons, and UAVs. The second method involves various sensors deployed for specific purposes at different sites. These collected data are then pinpointed with precise location information using GPS devices, allowing for tailored, site-specific treatments (Zhang et al., 2018).

In recent years, agriculture has transitioned from small- and medium-sized farms to a highly industrialized and commercial sector. This transformation allows major corporations to apply the same focus on measurement, data, and control as seen in manufacturing, with the goal of balancing costs and production to maximize profits. Consequently, IoT technologies and solutions are increasingly being used to automate, digitally plan, and manage various agricultural processes. Current efforts are focused on developing advanced tools, such as agricultural robots, to perform tasks like weeding, harvesting, planting, spraying, watering, thinning, fertilizing, packing, and transporting (Buyya & Dastjerdi, 2016). This shift is driven by not only technological advancements but also by factors such as the decreasing availability of low-cost labor and the growing demand for more affordable, high-quality food. Manufacturers in the market offer a wide array of products and solutions, mainly leveraging sensors and advanced communication technologies for a variety of applications. The key technologies and equipment available for these purposes are outlined below.

3.1. Sensors

Based on different operating principles, sensors in smart agriculture can be grouped into different categories. Based on measurement principles, sensors are listed below.

3.1.1. Optical Sensors (Murray et al., 2018)

Photometers: These sensors measure the amount of light that plants receive. They measure the intensity of light and help ensure that plants are getting the proper conditions for their growth.

Spectral sensors: These sensors use different wavelengths of light to assess the health of plants by measuring stress level and nutrient values. Key examples include Normalized Difference Vegetation Index sensors, multispectral sensors, and hyperspectral sensors, that offer more in-depth analysis.

Chlorophyll sensors: These sensors measure the amount of light absorption and reflection and thus determine the chlorophyll levels in plants. This information is important for evaluating the plant health, as the amount of chlorophyll is directly proportional to the plant's ability for effective photosynthesis and growth.

3.1.2. Electrochemical Sensors (Yew et al., 2014)

pH Sensors: They are designed to measure the acidity or alkalinity of water and soil by electrochemical methods. Checking pH levels is important for the purpose of improving soil conditions and ensuring that plants have access to nutrients.

Soil nutrient sensors: These sensors use electrochemical reactions to measure key nutrient levels, such as nitrogen, phosphorus, and potassium. This information helps in controlling soil fertility and managing fertilization.

Dissolved oxygen sensors: These sensors are designed to measure the amount of oxygen present in water. They help in maintaining healthy water conditions in irrigation systems and aquaculture suitable for both plant and animal life.

3.1.3. Capacitive Sensors

Soil moisture sensors: The dielectric constant of the soil changes with the amount of moisture it contains; these sensors are designed to help in measuring the soil's water levels. This information helps for optimizing the irrigation process, ensuring that plants receive the right amount of water for growth.

3.1.4. Resistive Sensors

Resistive soil moisture sensors: These sensors work on the principle of measuring the electrical resistance between two electrodes, which changes with respect to the moisture content in the soil. These data help in managing soil water levels, ensuring proper irrigation and plant health.

3.1.5. Thermal Sensors

Infrared sensors: By detecting heat emitted by plants or animals, these sensors are used in thermal imaging to find signs of stress, disease, or temperature change. This helps in the early detection of any problem associated with crops and livestock.

Temperature sensors: Thermocouple or thermistors are active sensors to measure ambient or soil temperature. They provide information for keeping ideal growing conditions and monitoring any environmental changes.

3.1.6. Ultrasonic Sensors (Dvorak et al., 2016)

Distance measurement sensors: These sensors use ultrasonic waves to measure the distance to objects, which make them best for jobs like determining plant height or assisting machinery in avoiding any obstacle.

3.1.7. Light Detection and Ranging (LIDAR) (Schuster et al., 2017)

LIDAR sensors: Unlike ultrasonic sensors, these sensors use laser pulses to determine distances and produce detailed three-dimensional maps about crop heights, field topography, and other important agricultural parameters.

3.1.8. Radio Frequency (RF) Sensors

Radio-Frequency Identification (RFID) tags: These tags use RF signals to track and identify animals or any equipment, making agricultural operations easier.

3.1.9. Mechanical Sensors (Hemmat et al., 2013)

Pressure sensors: These sensors are used to measure soil or water pressure and maintain water flow and are commonly used in irrigation systems for efficient and effective water management.

3.1.10. Conductivity Sensors

Electrical conductivity sensors: These sensors measure the soil's electrical conductivity, providing information about soil salinity and its nutrient levels, which are important for crop nutrition.

3.1.11. Chemical Sensors

Turbidity sensors: These sensors are used to measure the presence of sediments or contaminants in water by measuring cloudiness.

3.1.12. Hydroponic Sensors (Nishimura et al., 2017)

Electrical conductivity sensors: These sensors ensure that nutrients are supplied at a desired amount for plant growth by measuring the conductivity of nutrient solutions in hydroponic systems (a method of cultivating plants without soil).

Each type of sensor is designed for a specific purpose, and their integration into smart agriculture systems enables efficient management of an array of agricultural parameters.

3.2. IoT-based Tractors

IoT-based tractors are novel farming equipment that have been equipped with IoT technology to increase efficiency and productivity in the agriculture sector. These tractors maximize farming operations by combining GPS, sensors, and communication technologies (Oliver, 2017). Following are the important features of IoT-based tractors.

3.2.1. Precision Farming

GPS guidance: IoT tractors use GPS technology for accurate field mapping and navigation and thus help in ensuring planting and harvesting.

Automated steering: This system self-guides the tractor along predefined paths, thereby helps in minimizing overlap and increasing efficiency.

3.2.2. Real-time Monitoring

Sensors and data collection: IoT tractors fitted with different sensors monitor various parameters such as soil conditions, fuel levels, and engine performance. The data collected in real time are sent to the control system for further processing.

Remote diagnostics: The analysis of real-time data makes remote diagnostics possible, thus eliminating the need for physical presence.

3.2.3. Enhanced Efficiency

Fuel management: The real-time tracking of fuel helps in optimizing the use of fuel which reduces the operating cost and its impact on the environment.

Maintenance alerts: IoT systems helps to avoid breakdown and increases the life span of a tractor by alerting in advance for any probable problems and maintenance.

3.2.4. Automation and Control

Remote operation: IoT tractors can be operated autonomously or semi-autonomously and help in improving flexibility and reducing manual labor.

Precision application: Automated systems guarantee correct usage of pesticides, fertilizers, and other inputs, reducing waste and increasing effectiveness.

3.2.5. Fleet Management

Tracking and coordination: Fleet management systems keep eye on the locations and conditions of numerous tractors, maximizing their utility and aiding coordination across large farms.

IoT-based tractors represent a major development in the field of sustainable agricultural technology, aiding farmers to run their business more effectively.

3.3. Harvesting Robots

Harvesting robots are modern advanced machinery designed to harvest crops automatically which helps in increasing productivity and reducing operating cost. The primary purpose of these robots is to swiftly and precisely complete the harvesting task (Bargoti & Underwood, 2017; Zhao et al., 2016; Zujevs et al., 2015). Following are the important features of harvesting robots.

3.3.1. Advanced Sensors and Vision Systems

Image recognition: Harvesting robots easily identify the ripe fruits and vegetables with the help of high-resolution cameras and advanced image recognition software installed in it for the purpose of accurate selection and picking.

Depth sensors: These sensors determine the location and size of crops and help in reducing damage and ensuring careful handling.

3.3.2. Precision Handling

Robotic arms: These arms or grippers gently handle and harvest fragile crops to avoid any sort of damage or bruising.

Adaptive grippers: Some robots come with adaptive grippers that can be adjusted to hold different sizes and shapes of crops effectively.

3.3.3. Automation and Efficiency

Speed and efficiency: Harvesting robots substantially increase the harvest rate when compared with manual labor and minimize errors making harvesting operations more productive and efficient.

3.3.4. Versatility

Crop types and field conditions: Harvesting robots are designed to handle a variety of crops including grains, fruits, and vegetables; they are built to work in different field conditions, such as uneven ground and variable weather.

Harvesting robots are revolutionizing the agriculture sector by increasing efficiency, productivity, and sustainability.

3.4. Communication Technologies

Several communication technologies are used in IoT-based agriculture to link devices, send data, and perform various other tasks (Hong & Hsieh, 2016; Jedermann et al., 2018; Mendez et al., 2011; Petajajarvi et al., 2016; Taskin et al., 2018). An overview of different communication technologies is provided in Table 1.1. Each technology has unique advantages and is chosen as per the specific needs of the IoT application in agriculture.

Table 1.1. Different Communication Technologies.

Technology	Range	Bandwidth	Use Cases	Advantage	Disadvantage
Cellular communication	Wide-range, suitable for long-distance communication	High bandwidth, supporting large volumes of data	Ideal for scenarios where devices are dispersed across vast areas	Provides widespread coverage, often with reliable connectivity	Can be costly, especially in rural areas where network infrastructure might be limited
Bluetooth	Short range, typically up to 100 meters	Moderate bandwidth, suitable for smaller data packets	Ideal for connecting devices that are in close proximity	Low power consumption and cost-effective	Limited range and lower data transfer rates
Zigbee	Medium range, up to 100 meters (328 feet) in open areas	Low to moderate bandwidth, suitable for small data packets	Perfect for establishing mesh networks in smart agriculture	Low power consumption, scalable, and good for battery-operated devices	Limited data transfer rates and range
LoRa (long range)	Long range, up to 15 kilometers (9 miles) in open areas	Low bandwidth, suitable for small data packets	Ideal for applications needing wide-area coverage and low data rates	Very long range and low power consumption	Limited data transfer rates, which can be a constraint for applications requiring high data throughput

Source: Authors' compilation.

3.5. Cloud Computing

Cloud computing is essential to smart agriculture, as it offers real-time data processing and complicated analysis (Choudhary et al., 2016; Terence & Geethanjali, 2020). Cloud computing helps in managing data, real-time processing, advanced analytics, and remote access. It improves decision-making, operational efficiency, and provides economical solutions, all of which support more productive and sustainable farming. Table 1.2 illustrates the use of cloud computing in smart agriculture.

Table 1.2. Cloud Computing Utilization in Smart Agriculture.

Features	**Utilization**
Data storage and management	*Scalability*: Cloud platforms offer virtually unlimited storage, crucial for handling the vast amounts of data generated by sensors, drones, and other IoT devices.
	Accessibility: Cloud-stored data can be accessed from any location, enabling farmers and stakeholders to review and analyze information in real-time or historically.
Real-time data processing	*Data aggregation*: Cloud computing provides a complete picture of farm conditions by bridging data from various sensors, weather stations, and satellite imagery.
	Real-time analysis: Cloud services facilitate real-time data processing, enabling farmers to make prompt decisions based on up-to-date conditions and forecasts.
Advanced analytics and machine learning	*Predictive analytics*: Cloud platforms use machine learning algorithms to forecast weather patterns, crop diseases, and appropriate harvesting times.
	Decision support: Advanced analytics tools assist in optimizing irrigation schedules, nutrient application, and pest control strategies, boosting productivity and sustainability.
Integration and interoperability	*Unified platform*: Cloud computing integrates diverse agricultural technologies and systems – such as IoT devices, drones, and farm management software – into a single platform, streamlining operations.
	Application Programming Interface (APIs) and services: Cloud platforms offer APIs and services that promote interoperability between various systems and devices, ensuring smooth data exchange and process automation.
Remote monitoring and control	*Accessibility*: Farmers can remotely oversee and manage farm operations using cloud-based dashboards and mobile applications, enhancing management efficiency and responsiveness.

Table 1.2. (*Continued*)

Features	Utilization
	Alerts and notifications: Cloud systems provide real-time alerts and notifications about critical conditions or issues, enabling prompt intervention and reducing potential losses.
Data security and backup	*Security*: Cloud providers implement strong security measures, such as encryption and access controls, to safeguard sensitive agricultural data from unauthorized access and breaches.
	Backup and recovery: In case of system failure, cloud services offer data recovery.
Cost-efficiency	*Pay-as-you-go*: Cloud computing uses a pay-as-you-go model, to only pay for the resources they actually use, potentially saving money.
	Reduced maintenance: Cloud services remove the need for maintaining and upgrading physical hardware, lowering operational costs and information technology overhead.
Collaboration and sharing	*Data sharing*: Cloud platforms enable seamless data sharing among farmers, researchers, and agricultural experts, fostering collaboration and knowledge exchange.
	Community insights: By analyzing aggregated data from various farms, cloud services offer valuable insights and trends that benefit the wider agricultural community.

Source: Authors' compilation.

4. UAVs in Agriculture

UAVs commonly known as drones, are slowly becoming an important tool for agriculture sector to increase productivity, and efficiency (Anthony et al., 2014; Bendig et al., 2012; Faiçal et al., 2017; Huang et al., 2009; Kurkute et al., 2018; Primicerio et al., 2012; Talaviya et al., 2020).

4.1. Crop Monitoring and Scouting

Visual and multispectral imaging: Farmers can identify the crop disease, monitor crop health, and assess the growth of crops with the help of drones fitted with high-resolution camera and advanced image recognition software installed in it.

Thermal imaging: Drones fitted with thermal sensors can identify irrigation problems and improve water management by identifying variations in soil moisture and temperature.

4.2. Precision Agriculture

Variable rate application: On the basis of data collected with the help of drones, the application of fertilizers, pesticides, and water can be applied more effectively. This helps in minimizing cost and reduces environmental impact.

Soil and field analysis: Detailed mapping of fields can be done by drones which helps in soil analysis and soil sampling. This methodology helps in choosing the most effective crop planting schedules.

4.3. Spraying and Spreading

Aerial spraying: Drones fitted with sprayers evenly apply pesticides, herbicides, and fertilizers to fields, improving overall coverage and reducing direct chemical exposure for workers.

Seed spreading: Drones can easily spread seeds over large geographical areas, helping in the process of reforestation and crop planting.

4.4. Irrigation Management

Monitoring water stress: Drones mounted with infrared and multispectral sensors are used to find water stress levels in plants. This helps farmers to reschedule their irrigation requirements.

4.5. Livestock Monitoring

Herd tracking: Drones can easily track the movement of livestock and also ensure their safety and health.

Pasture management: Drones can estimate pasture quality that helps in managing grazing patterns more effectively.

4.6. Data Collection and Analysis

Integration with software: Drone data collected can be integrated with farm management software to provide more comprehensive maps and detailed analyses.

As technological advancement makes UAVs more reasonably available, their use in agriculture is continuously growing.

5. Challenges in IoT Adoption for Traditional Farmers

Precision agriculture is a valuable technology designed to address the growing demand for limited resources. Despite its benefits, there are several challenges that agriculturalists, farmers, researchers, and scientists encounter when adopting this technology. To understand regional perspectives on these challenges, a local survey was conducted to gather insights from the community.

5.1. Result and Discussion

Site description: Rajasthan, located in north western India, is the largest state by area and the seventh largest by population. Its geographical coordinates range from 23°.3' to 30°.12' North latitude and 69°.30' to 78°.17' East longitude, with the Tropic of Cancer passing through its southernmost tip. The state's winter temperatures range from 8° to 28°C, while summer temperatures range from 25° to 46°C. Rainfall varies significantly across the state, from 299.2 mm in the western regions to 671.3 mm in the eastern regions. With a population of 6.86 crore and a literacy rate of 66%, approximately 75% of Rajasthan's residents live in rural areas, and around 62% rely on agriculture and related activities for their livelihood. Agriculture in Rajasthan is predominantly rain-fed with a brief monsoon period, making farmers heavily dependent on both rainfall and groundwater for cultivation. Major kharif crops include groundnut, bajra, and kharif pulses, while rabi crops consist of wheat, mustard, barley, and gram. Common vegetables cultivated include tomato, pea, chili, brinjal, cabbage, and cauliflower.

To understand agricultural practices in the region, a local survey was conducted in the Jaipur district, targeting farmers ranging from small scale to large scale, who generally use simple technologies and traditional methods in their farming practice for agricultural information to improve productivity. A total of 270 farmers from four different agro-ecological situations (AES) in the district participated in the survey, aimed at identifying gaps in knowledge and skills. The survey involved interviews to understand the reasoning behind their agricultural practices. The survey's details, including the sites, number of respondents, and their average ages, are summarized in Table 1.3.

In this survey, various questions were posed to the farmers, who largely comprised uneducated individuals, though some had formal education and a few were knowledgeable about smart agriculture and modern technologies. Table 1.4 presents the different questions asked, along with the percentage of responses categorized as Yes or No. Positive responses were predominantly received from farmers who were either educated or had their fields under study.

Based on the results, it can be concluded that integrating technology into traditional farming practices has the potential to deliver substantial benefits. However, as highlighted by the local survey conducted in the Jaipur district, farmers may encounter several challenges such as high initial cost, lack of technical knowledge, connectivity issues, and resistance to change.

Table 1.3. Survey of Farmers.

Survey Site	No. of Respondents	Average Age
AES-I	63	49
AES-II	72	51
AES-III	91	47
AES-IV	44	54

Source: Authors' findings.

Table 1.4. Respondent's Profile.

Question Asked to the Farmers, Do You Have Knowledge On	Yes %	No %
Yield monitoring	4	96
Soil mapping	6	94
Variable rate technology	2	98
Smart agricultural practices	7	93
Data analytics in smart agriculture	5	95
IoT	6	94

Source: Authors' findings.

6. Conclusion, Limitation, and Future Research Direction

To address the growing food demand of a rapidly increasing global population amid shrinking arable land, it is crucial to focus on smarter, more efficient crop-growing methodologies. This chapter explores the importance of advanced crop disease management strategies and emphasizes the role of technologies, particularly IoT, in enhancing agricultural efficiency and meeting future demands. It provides an in-depth discussion on wireless sensors, UAVs, cloud computing, and communication technologies, as well as recent research efforts. Additionally, it examines various IoT-based architectures and platforms relevant to agricultural applications. This chapter also addresses the challenges encountered in developing and deploying these technologies in real-world scenarios, including insights from a local survey on perceptions of precision agriculture. Despite the potential benefits, significant challenges remain for large-scale implementation of these frameworks.

There are a few limitations to the current study that future studies should acknowledge. First, the sample consisted of respondents from a specific geographical region. A larger sample from different geographical locations could lead to a different picture of smart farming. Second, this study was limited to specific technology and usage.

Further, the implementation of sustainable policies in IoT provides solutions to the farmers to encounter several challenges as already discussed. These challenges underscore the necessity for training and support systems to enable traditional farmers to effectively adopt smart practices and get benefited from new technologies.

References

Abdelrahman, M. A. E. (2023). An overview of land degradation, desertification and sustainable land management using GIS and remote sensing applications. *Rendiconti Lincei Scienze Fisiche e Naturali*, *34*, 767–808.

Anthony, D., Elbaum, S., Lorenz, A., & Detweiler, C. (2014, September 14–18). *On crop height estimation with UAVs* (pp. 4805–4812). IROS 2014 Conference Digest, IEEE/

RSJ International Conference on Intelligent Robots and Systems, Chicago. https://doi.org/10.1109/IROS.2014.6943245

Ayaz, M., Ammad-Uddin, M., Sharif, Z., Mansour, A., & Aggoune, E. H. M. (2019). Internet-of-things (IoT)-based smart agriculture: Toward making the fields talk. *IEEE Access*, *7*, 129551–129583.

Bargoti, S., & Underwood, J. (2017, May 29–June 3). *Deep fruit detection in orchards* (pp. 3626–3633). Proceedings of the 2017 IEEE International Conference on Robotics and Automation (ICRA), Marina Bay Sands, Singapore.

Bendig, J., Bolten, A., & Bareth, G. (2012). *Introducing a low-cost mini-UAV for thermal- and multispectral-imaging* (XXXIX-B1, pp. 345–349). ISPRS – International Archives of the Photogrammetry, Remote Sensing and Spatial Information Sciences, Copernicus Publications.

Bendre, M., Thool, R. C., & Thool, V. R. (2015). *Big data in precision agriculture: Weather forecasting for future farming* (pp. 744–750). 2015 1st international conference on next generation computing technologies (NGCT).

Bhargava, K., Kashyap, A., & Gonsalves, T. A. (2014). Wireless sensor network based advisory system for Apple Scab prevention. In *2014 20th National Conference on Communications, NCC 2014* (pp. 1–6). IEEE Computer Society. https://doi.org/10.1109/NCC.2014.6811263

Buyya, R., & Dastjerdi, A. V. (2016). *Internet of things: Principles and paradigms*.Elsevier.

Carvalho, F. P. (2017). Pesticides, environment, and food safety. *Food and Energy Security*, *6*, 48–60. https://doi.org/10.1002/fes3.108

Choudhary, S. K., Jadoun, R. S., & Mandoriya, H. L. (2016). Role of cloud computing technology in agriculture fields. *Computer Engineering and Intelligent Systems*, *7*, 1–7.

Cruz-Blanco, M, Lorite, I. J., & Santos, C. (2014). An innovative remote sensing based reference evapotranspiration method to support irrigation water management under semi-arid conditions. *Agricultural Water Management*, *131*(C), 135–145.

Dvorak, J. S., Stone, M. L., & Self, K. P. (2016). Object detection for agricultural and construction environments using an ultrasonic sensor. *Journal of Agricultural Safety and Health*, *22*(2), 107–119. https://doi.org/10.13031/jash.22.11260

Elijah, O., Rahman, T. A., Orikumhi, I., Leow, C. Y., & Hindia, M. N. (2018). An overview of internet of things (IoT) and data analytics in agriculture: Benefits and challenges. *IEEE Internet Things Journal*, *5*(5), 3758–3773.

Faiçal, B. S., Freitas, H., Gomes, P. H., Mano, L. Y., Pessin, G., de Carvalho, A. C. P. L. F., Krishnamachar, B., & Ueyama, J. (2017). An adaptive approach for UAV-based pesticide spraying in dynamic environments. *Computers and Electronics in Agriculture*, *138*, 210–223.

Farooq, M. S., Riaz, S., Abid, A., Umer, T., & Zikria, Y. B. (2020). Role of IoT technology in agriculture: A systematic literature review. *Electronics*, *9*(2), 1–41.

Food and Agriculture Organization. (2019). *Pest and pesticide management*. https://www.fao.org/pest-and-pesticide-management/about/understanding-the-context/en/

Gulati, A., & Juneja, R. (2022). Transforming Indian agriculture. In R. Chand, P., Joshi, & S. Khadka (Eds.), *Indian agriculture towards 2030. India studies in business and economics*. Springer. https://doi.org/10.1007/978-981-19-0763-0_2

Han, P., Dong, D., Zhao, X., Jiao, L., & Lang, Y. (2016). A smartphone-based soil color sensor: For soil type classification. *Computers and Electronics in Agriculture*, *123*, 232–241.

Hemmat, A., Binandeh, A. R., Ghaisari, J., & Khorsandi, A. (2013). Development and field testing of an integrated sensor for on-the-go measurement of soil mechanical resistance. *Sensors and Actuators A: Physical*, *198*, 61–68.

Hi, X., An, X., Zhao, Q., Liu, H., Xia, L., Sun, X., & Guo, Y. (2019). State-of-the-art internet of things in protected agriculture. *Sensors*, *19*(8), 1–24.

Hong, G. Z., & Hsieh, C. L. (2016). Application of integrated control strategy and Bluetooth for irrigating romaine lettuce in greenhouse. *IFAC-Papers Online*, *49*(16), 381–386.

Hoogeveen, M., Ono, Y., Pfister, S., Boulay, A. M., Berger, M., Nansai, K., Tahara, K., Itsubo, N., & Inaba, A. (2018). Consistent characterisation factors at midpoint and endpoint relevant to agricultural water scarcity arising from freshwater consumption. *International Journal of Life Cycle Assessment*, *23*(12), 2276–2287.

Huang, Y., Hoffmann, W. C., Lan, Y., Wu, W., & Fritz, B. K. (2009). Development of a spray system for an unmanned aerial vehicle platform. *Applied Engineering in Agriculture*, *25*(6), 803–809.

Jawahar, I., Mohamed, N., Jaroodi, J. A., & Zhang, S. (2014). A framework for using unmanned aerial vehicles for data collection in linear wireless sensor networks. *Journal of Intelligent & Robotic Systems*, *74*, 437–453.

Jedermann, R., Borysov, M., Hartgenbusch, N., Jaeger, S., Sellwig, M., & Lang, W. (2018). Testing Lora for food applications – Example application for air flow measurements inside cooled warehouses with apples. *Procedia Manufacturing*, *24*, 284–289.

Kumar, A., Rani, M., Aishwarya, & Kumar, P. (2022). Drone technology in sustainable agriculture: The future of farming is precision agriculture and mapping. In A. Kumar, P. Kumar, S. S. Singh, B. H. Trisasongko, & M. Rani (Eds.), *Agriculture, livestock production and aquaculture* (pp. 3–12). Springer.

Kumar, R., Mishra, R., Gupta, H. P., & Dutta, T. (2021). Smart sensing for agriculture: Applications, advancements, and challenges. *IEEE Consumer Electronics Magazine*, *10*, 51–56.

Kurkute, S. R., Deore, B. D., Kasar., P, Bhamare., M., & Mayuri, S. (2018). Drones for smart agriculture: A technical report. *International Journal for Research in Applied Science & Engineering Technology*, *6*(4), 341–346.

Lin, J., Yu, W., Zhang, N., Yang, X., Zhang, H., & Zhao, W. (2017). A survey on Internet of things: Architecture, enabling technologies, security and privacy, and applications. *IEEE Internet Things Journal*, *4*(5), 1125–1142.

Liu, H., Wang, X., & Bing-Kun, J. (2018). Study on NDVI optimization of corn variable fertilizer applicator. *Agricultural Engineering*, *56*(3), 193–202.

Luigi, L. M., Emanuele, E. P., Zibordi, M., Morandi, B., Muzzi, E., Losciale, P., Corelli, L., & Grappadelli, L. C. (2015). Monitoring strategies for precise production of high quality fruit and yield in apple in Emilia Romagna. *Chemical Engineering Transactions*, *44*, 301–306.

Mendez, G. R., Yunus, M. A. M., & Mukhopadhyay, S. C. (2011). *A W-Fi based smart wireless sensor network for an agricultural environment* (pp. 405–410). 5th international conference on sensors and technology.

Mohamed, E. S., Belal, A. A., Elmabod, S. K., El-Shirbeny, M. A., Gad, A., & Zahran, M. B. (2021). Smart farming for improving agricultural management. *The Egyptian Journal of Remote Sensing and Space Sciences*, *24*, 971–981.

Murray, S. C. (2018). Optical sensors advancing precision in agricultural production. *Photonics Spectra*, *51*, 48.

Musie, W., & Gonfa, G. (2023). Fresh water resource, scarcity, water salinity challenges and possible remedies: A review. *Heliyon*, *9*(8). https://doi.org/10.1016/j.heliyon.2023.e18685

Navulur, S., & Prasad, M. N. G. (2017). Agricultural management through wireless sensors and internet of things. *International Journal of Electrical and Computing Engineering*, *7*(6), 3492–3499.

Nishimura, T., Okuyama, Y., Matsushita, A., Ikeda, H., & Satoh, A. (2017). *A compact hardware design of a sensor module for hydroponics* (pp. 1–4). IEEE 6th global conference on consumer electronics (GCCE), Nagoya.

Oberti, R., Marchi, M., Tirelli, P., Calcante, A., Iriti, M., Tona, E., Hocevar, M., Baur, J., Pfaff, J., Schutz, C., & Ulbrich, H. (2016). Selective spraying of grapevines for disease control using a modular agricultural robot. *Biosystems Engineering*, *146*, 203–215.

Oliveira, K. V., Castelli, H. M. E., Montebeller, S. J., & Avancini, T. G. P (2017). *Wireless sensor network for smart agriculture using ZigBee protocol* (pp. 61–66). IEEE 1st summer school smart cities (S3C), Natal, Brazil.

Oliver, J. (2017, August). *e-Agriculture – Internet of things (IoT) for agriculture webinar series: IoT: The internet of tractors*. http://www.fao.org/e-agriculture/news/internet-things-lot-agriculture-webinar-series-iot-internet-tractors

Petajajarvi, J., Mikhaylov, K., Hamalainen, M., & Iinatti, J. (2016). *Evaluation of LoRa LPWAN technology for remote health and wellbeing monitoring* (pp. 1–5). 10th international symposium on medical information and communication technology (ISMICT), Worcester, MA, USA.

Primicerio, J., Di Gennaro, S. F., Fiorillo, E., Genesio, L., Lugato, E., Matese, A., & Vaccari, F. P. (2012). A flexible unmanned aerial vehicle for precision agriculture. *Precision Agriculture*, *13*(4), 517–523.

Schuster, J. N., Darr, M. J., & McNaull, R. P. (2017). *Performance benchmark of yield monitors for mechanical and environmental influences*. Agricultural and biosystems engineering conference proceedings and presentations, IOWA State University, Ames, IA, USA.

Shinde, S., & Kulkarni, M. (2017). *Review paper on prediction of crop disease using IoT and machine learning* (pp. 1–4). 2017 International conference on transforming engineering education (ICTEE).

Sisinni, E., Saifullah, A., Han, S., Jennehag, U., & Gidlund, M. (2018). Industrial internet of things: Challenges, opportunities, and directions. *IEEE Transactions on Industrial Informatics*, *14*(11), 4724–4734.

Talaviya, T., Shah, D., Patel, N., Yagnik, H., & Shah, M. (2020). Implementation of artificial intelligence in agriculture for optimisation of irrigation and application of pesticides and herbicides. *Artificial Intelligence in Agriculture*, *4*, 58–73.

Taskin, D., Taskin, C., & Yazar, S. (2018). Developing a Bluetooth low energy sensor node for greenhouse in precision agriculture as internet of things application. *Advances in Science Technology and Research Journal*, *12*(4), 88–96.

Terence, S., & Geethanjali, P. (2020). Systematic review of internet of things in smart farming. *Transactions on Emerging Telecommunications Technologies*, *31*(6).

Tripathi, A. D., Mishra, R., Maurya, K. K., Singh, R. B., & Wilson, D. W. (2019). Estimates for world population and global food availability for global health. In R. B. Singh, R. R. Watson, & T. Takahashi (Eds.), *The role of functional food security in global health* (pp. 3–24). Elsevier.

Universe Today. (2014). https://www.universetoday.com/65588/what-percent-of-earth-is-water/#google_vignette

Waskom, R., Bauder, T., Pearson, R. (2017). *Best management practices for agricultural pesticide use* (Bulletin #XCM-177). Colorado State University Cooperative Extension.

Yew, T. K., Yusoff, Y., Sieng, L. K., Lah, H. C., Majid, H., & Shelida, N. (2014). *An electrochemical sensor ASIC for agriculture applications* (pp. 85–90). 37th international convention on information and communication technology, electronics and microelectronics (MIPRO), Opatija, Croatia.

Zhang, L., Dabipi, I. K., & Brown, W. L. (2018). Internet of things applications for agriculture. In Q. Hassan (Ed.), *Internet of things A to Z: Technologies and applications* (pp. 507–528). Wiley.

Zhao, Y., Gong, L., Huang, Y., & Liu, C. (2016). A review of key techniques of vision-based control for harvesting robot. *Computers and Electronics in Agriculture*, *127*, 311–323.

Zujevs, A., Osadcuks, V., & Ahrendt, P. (2015). Trends in robotic sensor technologies for fruit harvesting: 2010–2015. *Procedia Computer Science*, *77*, 227–233.

Chapter 2

Bridging the Knowledge Gap: How Higher Education Can Foster Adaptive Management in Crop Ecosystems

Eutychus Ngotho Gichuru

College of Education and External Studies, Makerere University, Kampala, Uganda

Abstract

By incorporating new ways of transmitting information and revamping the curriculum, higher education institutions play an important role in nurturing adaptive management (AM) in crop ecosystems. Higher education institutions are very important when it comes to fostering AM in crop ecosystems through innovative knowledge-sharing approaches and curriculum development. The COVID-19 pandemic has hastened the transition to online learning revealing how crucial optimal knowledge sharing is for educational quality and accessibility that should not be forgotten again. These steps include transdisciplinary practices, resources mobilization, and leveraging upon partnerships to address challenges that can drive sustainable transformation in agriculture and food systems. In this line, such communities of practice may be established to serve as platforms for implementing decentralized education systems, thereby refining students' learning experiences, thus enhancing program quality improvement as well as specialization on specific fields which significantly contributes to AM within the crop ecosystems. An approach to educational construction which is systemic and interdisciplinary can address knowledge gaps and prepare future practitioners to cope with changing complexities of crop ecosystems using adaptive approaches and innovative solutions, thereby promoting sustainable agricultural practices and better ecosystem management.

Keywords: Adaptive management; curriculum development; decentralized education; higher education institutions; knowledge sharing

Sustainable and Smart Farming Eco-systems:
Changing Landscapes in the Age of Technology, 21–30

doi:10.1108/978-1-83608-092-320261002

Introduction

AM has emerged as a crucial framework for addressing uncertainty in various sectors, including higher education and agriculture. Higher education institutions have successfully embraced AM by integrating iterative learning processes, stakeholder participation, and evidence-based decision-making into their management practices. Higher education institutions have recognized the importance of AM in navigating the complexities of educational management. As noted by Bryan et al. (2009), successful applications of AM are rare, yet higher education has demonstrated a commitment to iterative learning and adaptation. By fostering a culture of continuous improvement, universities can respond effectively to changing educational needs and societal demands.

One of the key components of AM is stakeholder participation, which is essential for managing conflicts and enhancing decision-making (Rist et al., 2013). Higher education institutions actively engage students, faculty, and community members in the decision-making process, ensuring that diverse perspectives are considered. This participatory approach not only improves management outcomes but also fosters a sense of ownership and accountability among stakeholders. In the context of AM, higher education institutions have also excelled in defining and bounding management problems, setting clear objectives, and representing existing knowledge through system models (Rist et al., 2013). By establishing well-defined goals and utilizing data-driven models, universities can effectively assess their performance and make informed decisions. This structured approach can serve as a model for the agricultural industry, which often grapples with complex challenges related to resource management and sustainability.

Monitoring and evaluation are critical components of AM, allowing institutions to reflect on their actions and learn from outcomes (Rist et al., 2013). Higher education institutions routinely assess the effectiveness of their programs and policies, using feedback to refine their strategies. The agricultural industry can adopt similar practices by implementing robust monitoring systems to evaluate the impact of management decisions on crop yields, resource use, and environmental sustainability.

Moreover, the iterative nature of AM encourages continuous learning and adaptation, which is vital for both higher education and agriculture (Rist et al., 2013). Universities often cycle through planning, implementation, and evaluation phases, allowing them to adjust their strategies based on new information and changing circumstances. The agricultural sector can benefit from this iterative approach by embracing flexibility and responsiveness in their management practices. Despite the successes of AM in higher education, challenges remain in its application. As highlighted by Eberhard et al. (2014), documented failure rates in AM practices are high. Higher education institutions must continue to address these challenges by fostering a culture of experimentation and learning from failures. The agricultural industry can similarly benefit from adopting a mindset that views setbacks as opportunities for growth and improvement.

The lessons learned from higher education's embrace of AM can significantly enhance the agricultural industry's capacity to adapt to changing conditions.

By prioritizing stakeholder engagement, establishing clear objectives, and implementing robust monitoring systems, the agricultural sector can improve its management practices and outcomes. Furthermore, adopting an iterative approach to decision-making will enable farmers and agricultural organizations to respond effectively to emerging challenges.

Higher education institutions have successfully embraced AM by integrating stakeholder participation, evidence-based decision-making, and iterative learning processes into their practices. The agricultural industry can learn from these experiences by adopting similar strategies to enhance its resilience and adaptability. By fostering a culture of continuous improvement and learning, both sectors can better navigate the complexities of their respective environments and achieve sustainable outcomes.

Higher education institutions play a pivotal role in nurturing AM within crop ecosystems by revamping curricula and adopting innovative knowledge-sharing approaches. The literature indicates that AM is essential for addressing the complexities and uncertainties inherent in agricultural practices (Walters & Hilborn, 1978). By fostering a culture of collaboration and information sharing, universities can enhance decision-making processes and drive innovations in the agricultural sector (Johnson, 1999).

The COVID-19 pandemic has accelerated the transition to online learning, underscoring the importance of effective knowledge sharing for educational quality and accessibility (Rist et al., 2013). This shift has revealed the potential of digital platforms to facilitate transdisciplinary practices and resource mobilization, which are crucial for addressing the challenges faced by agriculture and food systems. Higher education institutions can leverage partnerships to create communities of practice that serve as platforms for decentralized education, ultimately refining students' learning experiences and enhancing program quality (Kallis et al., 2009).

Transdisciplinary approaches in higher education can significantly contribute to AM in crop ecosystems. By integrating diverse fields of study, universities can prepare future practitioners to cope with the complexities of agricultural systems (Ogden & Innes, 2009). This systemic and interdisciplinary educational framework can help bridge knowledge gaps and equip students with the necessary skills to implement adaptive solutions in agriculture (Ladson & Argent, 2002).

Moreover, the establishment of intelligent learning environments can enhance the convergence of educational models, allowing for personalized training trajectories and continuous program updates (Kay & Alder, 2017). Such environments foster adaptive decision-making skills that are vital for effective crop ecosystem management. The literature emphasizes the need for educational institutions to create ecosystems that enable students to access courses from various institutions, promoting a decentralized education system that supports AM practices (Walters, 1997).

The integration of AM principles into higher education curricula can also address the wicked problems associated with food systems and climate change (Holling & Meffe, 1996). By emphasizing multidisciplinary cooperation and collaborative knowledge construction, universities can prepare students to tackle the intricate dynamics of crop ecosystems. This approach aligns with the findings of Karkkainen (2002), who advocates for adaptive ecosystem management as a means to enhance resilience in natural resource management.

Furthermore, the literature suggests that fostering a culture of sharing and collaboration within higher education can lead to improved access to information and resources, ultimately driving innovation in agriculture (Wasserberg et al., 2009). By creating networks of knowledge exchange, universities can facilitate the dissemination of best practices and adaptive strategies among agricultural practitioners. This collaborative approach is essential for developing sustainable agricultural practices and better ecosystem management.

Higher education institutions have a critical role in fostering AM in crop ecosystems through innovative knowledge-sharing approaches and curriculum development. By embracing transdisciplinary practices, leveraging partnerships, and creating intelligent learning environments, universities can equip future professionals with the skills necessary to navigate the complexities of agricultural systems. The literature highlights the importance of a systemic and interdisciplinary educational framework in addressing the challenges of food systems and climate change, ultimately contributing to sustainable agricultural practices and improved ecosystem management.

Adaptive Management

AM is increasingly recognized as a crucial approach for enhancing crop ecosystems, particularly through the innovative practices adopted by higher education institutions. By revising curricula and integrating new methods of information dissemination, universities can significantly contribute to fostering AM in agricultural practices. This critical review of the literature explores how higher education can advance AM through knowledge-sharing initiatives, curriculum development, and the creation of intelligent learning environments.

Higher education institutions play a central role in promoting AM in crop ecosystems by facilitating innovative knowledge-sharing approaches. As Rist et al. (2013) highlight, stakeholder participation and collaboration are essential elements of AM. By nurturing a culture of information exchange, universities can improve decision-making processes and expand access to information, ultimately driving innovation in agricultural practices.

The COVID-19 pandemic has expedited the shift to online learning, highlighting the importance of effective knowledge sharing for both educational quality and accessibility. This transition has underscored the potential for universities to utilize digital platforms to distribute information and encourage collaboration among students and professionals in the agricultural sector. These technological advancements can support the development of AM practices that are responsive to the evolving challenges faced by crop ecosystems.

Transdisciplinary approaches and resource mobilization are vital for tackling the complex issues within agriculture and food systems (Rist et al., 2013). Higher education institutions can create communities of practice that act as platforms for decentralized education systems, refining the learning experiences of students and enhancing program quality. By promoting cross-disciplinary collaboration, universities can equip future agricultural professionals with the skills necessary to implement AM strategies effectively.

The establishment of intelligent learning environments can further strengthen the integration of educational models and personalized training trajectories (Rist et al., 2013). These environments support continuous program updates and the development of adaptive decision-making skills, which are key for managing crop ecosystems. By incorporating technology and innovative teaching methods, universities can establish a more responsive educational framework that prepares students for the complexities of agricultural management.

A systemic, interdisciplinary approach to education is essential for bridging knowledge gaps and preparing future professionals to address the changing dynamics of crop ecosystems (Rist et al., 2013). By emphasizing multidisciplinary collaboration and joint knowledge creation, higher education institutions can equip students with the tools necessary to address the complex challenges related to food systems and climate change. This approach aligns with the principles of AM, which emphasize learning and flexibility in decision-making.

Additionally, the promotion of decentralized education systems can improve access to a wide range of courses and expertise across different institutions (Rist et al., 2013). This flexibility allows students to customize their educational journeys based on personal interests and career aspirations, fostering a more adaptive mindset. By encouraging students to engage with diverse perspectives and practices, universities can cultivate a new generation of agricultural professionals capable of implementing AM strategies.

Higher education institutions play an essential role in advancing AM in crop ecosystems by facilitating innovative knowledge-sharing practices and developing curricula. Through fostering collaboration, creating intelligent learning environments, and promoting interdisciplinary approaches, universities can prepare future practitioners to navigate the complexities of agricultural management. As the literature suggests, integrating AM principles into higher education can lead to sustainable agricultural practices and improved ecosystem management.

Curriculum Development

By incorporating innovative methods of information dissemination and revising educational structures, universities can cultivate AM practices that foster enhanced decision-making, collaboration, and knowledge exchange. This review of the literature explores the significant contributions that curriculum development in higher education can make toward the promotion of AM in agricultural systems.

Higher education institutions are central to the promotion of AM in crop ecosystems, particularly through the cultivation of a culture of collaboration and information sharing. As Rist et al. (2013) observe, stakeholder participation is a core element of AM, and universities can facilitate this by designing curricula that prioritize collaborative learning and knowledge exchange. By integrating transdisciplinary approaches into academic programs, institutions can better prepare students to navigate complex, multifaceted agricultural challenges, ultimately leading to more innovative and context-specific solutions in the agricultural sector.

In rethinking their curricula, universities should consider adopting new methods of information transmission, including digital platforms and virtual collaboration tools. These innovations not only enhance access to educational resources but also contribute to the creation of more inclusive learning environments, thereby better equipping students to meet the demands of AM in agriculture. The integration of these technologies can also foster more dynamic and flexible learning processes, which are essential for responding to the evolving nature of agricultural practices and environmental conditions.

Communities of practice serve as valuable platforms for the implementation of decentralized education systems, refining student learning experiences and enhancing program quality (Rist et al., 2013). By establishing networks that connect students, faculty, and industry professionals, universities can facilitate the exchange of best practices and critical knowledge, which is vital for AM. Such collaborative structures not only promote shared learning but also allow for specialization in specific areas of agricultural science, equipping students with the expertise necessary to address the complex challenges facing crop ecosystems.

Furthermore, intelligent learning environments offer the potential to enhance the convergence of educational models and individualized learning trajectories (Rist et al., 2013). Through the integration of technology and innovative pedagogical methods, universities can design adaptive learning experiences that engage students in deeper, more meaningful interactions with course content. This approach enables the development of adaptive decision-making skills, which are crucial for effective management of crop ecosystems, as it supports students in learning how to make informed, flexible decisions in response to dynamic environmental and agricultural challenges.

A systemic and interdisciplinary approach to curriculum development is fundamental for addressing the existing knowledge gaps and preparing future practitioners to navigate the changing dynamics of crop ecosystems (Rist et al., 2013). By emphasizing cross-disciplinary collaboration and the joint construction of knowledge, higher education institutions can equip students with the necessary tools to tackle the complex, "wicked" problems associated with food systems and climate change. This approach aligns with the core principles of AM, which prioritize continuous learning and flexible, context-driven decision-making.

Such flexibility allows students to tailor their academic trajectories to their specific interests and professional goals, fostering a mindset that is adaptable and responsive to the changing demands of the agricultural sector. By encouraging engagement with diverse perspectives and practices, universities can better prepare students to implement AM strategies effectively in a variety of agricultural contexts.

Decentralized Education

Decentralized education systems provide increased flexibility and accessibility in learning, which are crucial for promoting AM in crop ecosystems. As Rist et al. (2013) emphasize, stakeholder participation is a foundational element of

AM, and decentralized educational models facilitate this by enabling students to engage with diverse perspectives and practices. By establishing systems that allow students to access courses across multiple institutions, universities can foster a culture of collaboration and knowledge sharing, thereby enhancing the principles of AM.

This transition has prompted universities to reconsider traditional educational paradigms and adopt decentralized models that utilize digital platforms for the dissemination of information. Such innovations not only improve access to educational resources but also cultivate a more inclusive learning environment, better preparing students to address the challenges of AM in agriculture.

Communities of practice represent an important mechanism for implementing decentralized education systems, enhancing student learning experiences, and improving program quality (Rist et al., 2013). By establishing networks that connect students, faculty, and industry professionals, universities can facilitate the exchange of best practices and critical knowledge, which is essential for AM. This collaborative framework enables specialization in particular areas of agriculture, thereby equipping students with the expertise necessary to navigate the complexities of crop ecosystems.

Through the use of technology and innovative pedagogical approaches, universities can create adaptive learning experiences that encourage students to engage more deeply with course material. This flexibility allows students to develop the adaptive decision-making skills required for effective management of crop ecosystems. By emphasizing cross-disciplinary cooperation and collaborative knowledge construction, higher education institutions can equip students with the tools necessary to tackle the complex issues associated with food systems and climate change. Such an approach aligns with the core principles of AM, which prioritize continuous learning and flexibility in decision-making.

This increased access allows students to tailor their educational paths to their specific interests and professional aspirations, ultimately nurturing a more adaptive mindset. By encouraging students to engage with diverse perspectives and practices, universities can better prepare them to implement AM strategies effectively.

Higher Education Institutions

Higher education institutions are uniquely positioned to support AM in crop ecosystems by promoting a culture of collaboration and knowledge exchange. As Rist et al. (2013) point out, stakeholder participation is a foundational element of AM, and universities can facilitate this by designing curricula that emphasize collaborative learning and the exchange of ideas across disciplines. By embedding transdisciplinary approaches within academic programs, institutions can enhance students' capacity to navigate complex agricultural challenges, ultimately fostering more innovative and holistic solutions in the field.

This shift has prompted universities to rethink traditional educational models and adopt decentralized approaches that leverage digital platforms and virtual collaboration tools for information dissemination. Such innovations not only broaden access to educational resources but also create a more inclusive and flexible learning environment, better equipping students to meet the challenges of AM in agriculture. By establishing networks that connect students, faculty, and industry professionals, universities can facilitate the exchange of best practices and critical knowledge – key components of AM. This collaborative approach also enables specialization in specific agricultural fields, equipping students with the expertise necessary to address the complex dynamics of crop ecosystems.

By leveraging cutting-edge technologies and innovative pedagogical strategies, universities can create adaptive learning experiences that enable students to engage with course content in deeper and more meaningful ways. This flexibility not only supports students in acquiring theoretical knowledge but also hones their adaptive decision-making skills – critical competencies for managing crop ecosystems effectively in a rapidly changing world.

By promoting multidisciplinary cooperation and fostering collaborative knowledge construction, higher education institutions can equip students with the tools necessary to confront the "wicked" problems of food systems and climate change. This approach aligns directly with the principles of AM, which emphasize learning, flexibility, and context-specific decision-making. This access enables students to tailor their educational journeys to align with their individual interests and career goals, ultimately fostering a more adaptive and responsive mindset. By encouraging students to engage with a variety of perspectives and practices, universities can better prepare them to implement AM strategies across diverse agricultural settings.

Knowledge Sharing

Universities have increasingly turned to digital platforms and virtual collaboration tools to facilitate the transmission of information. These technological innovations not only enhance access to educational resources but also create more inclusive learning environments, better preparing students to tackle the challenges of AM in agriculture. By leveraging these technologies, universities can create more dynamic learning experiences, allowing students to engage with course material in deeper and more meaningful ways. Such flexibility cultivates adaptive decision-making skills, which are essential for effective crop ecosystem management in an ever-changing agricultural landscape.

By emphasizing cross-disciplinary cooperation and fostering collaborative knowledge construction, higher education institutions can provide students with the tools necessary to confront the complex challenges posed by food systems and climate change. This approach aligns directly with the core principles of AM, which emphasize flexibility, continuous learning, and context-specific decision-making.

Conclusion

Higher education institutions are pivotal in fostering AM within crop ecosystems by implementing innovative knowledge-sharing practices and revamping curricula. By promoting a culture of collaboration and utilizing digital platforms, universities can enhance decision-making processes, improve access to information, and ultimately drive innovations in agricultural practices. The integration of transdisciplinary approaches and decentralized education systems equips future agricultural professionals with the necessary skills to navigate the complexities of crop ecosystems, thereby contributing to sustainable agricultural practices and improved ecosystem management.

Recommendations

To further enhance AM in crop ecosystems, higher education institutions should prioritize the establishment of communities of practice that facilitate knowledge sharing among students, faculty, and industry professionals. These communities can serve as platforms for decentralized education systems, allowing for the exchange of best practices and innovative solutions tailored to local agricultural challenges. By fostering collaboration across disciplines and creating networks of knowledge exchange, universities can better prepare students to tackle the intricate dynamics of crop ecosystems and promote sustainable agricultural practices.

Additionally, universities should leverage technology to create intelligent learning environments that support personalized training pathways and continuous program updates. By integrating adaptive learning experiences into their curricula, institutions can cultivate adaptive decision-making skills among students, which are essential for effective crop ecosystem management. Emphasizing a systemic and interdisciplinary approach will help bridge knowledge gaps and equip future practitioners with the tools needed to address the wicked problems associated with food systems and climate change. By adopting these strategies, higher education institutions can significantly contribute to the resilience and adaptability of the agricultural sector.

References

Bryan, B. A., Kandulu, J., Deere, D. A., White, M., Frizenschaf, J., & Crossman, N. D. (2009). Adaptive management for mitigating Cryptosporidium risk in source water: A case study in an agricultural catchment in South Australia. *Journal of Environmental Management*, *90*(10), 3122–3134. https://doi.org/10.1016/j.jenvman.2009.05.014

Eberhard, A., Kolker, J., & Leigland, J. (2014). *South Africa's renewable energy IPP procurement program: Success factors and lessons*. World Bank Group.

Holling, C. S., & Meffe, G. K. (1996). Command and control and the pathology of natural resource management. *Conservation Biology*, *10*(2), 328–337. https://doi.org/10.1046/j.1523-1739.1996.10020328.x

Johnson, B. L. (1999). Introduction to the special feature: Adaptive management – Scientifically sound, socially challenged? *Conservation Ecology*, *3*(1). https://doi.org/10.5751/es-00094-030110.

Kallis, G., Kiparsky, M., & Norgaard, R. (2009). Collaborative governance and adaptive management: Lessons from California's CALFED Water Program. *Environmental Science & Policy*, *12*(6), 631–643. https://doi.org/10.1016/j.envsci.2009.07.002

Karkkainen, B. C. (2002). Adaptive ecosystem management and regulatory penalty defaults: Toward a bounded pragmatism. *Minnesota Law Review*, *87*, 943. https://doi.org/10.24926/265535.1660

Kay, R., & Alder, J. (2017). *Coastal planning and management*. CRC Press. https://doi.org/10.1201/9781315272634-5

Ladson, A. R., & Argent, R. M. (2002). Adaptive management of environmental flows: Lessons for the Murray-Darling Basin from three large North American rivers. *Australasian Journal of Water Resources*, *5*(1), 89–101. https://doi.org/10.1080/13241583.2002.11465195

Ogden, A. E., & Innes, J. L. (2009). Application of structured decision making to an assessment of climate change vulnerabilities and adaptation options for sustainable forest management. *Ecology and Society*, *14*(1). https://doi.org/10.5751/es-02771-140111

Rist, L., Campbell, B. M., & Frost, P. (2013). Adaptive management: Where are we now? *Environmental Conservation*, *40*(1), 5–18. https://doi.org/10.1017/s0376892912000240

Walters, C. (1997). Challenges in adaptive management of riparian and coastal ecosystems. *Conservation Ecology*, *1*(2), 1–13. https://doi.org/10.5751/es-00026-010201

Walters, C. J., & Hilborn, R. (1978). Ecological optimization and adaptive management. *Annual Review of Ecology and Systematics*, *9*(1), 157–188. https://doi.org/10.1146/annurev.es.09.110178.001105

Wasserberg, G., Osnas, E. E., Rolley, R. E., & Samuel, M. D. (2009). Host culling as an adaptive management tool for chronic wasting disease in white-tailed deer: A modelling study. *Journal of Applied Ecology*, *46*(2), 457–466. https://doi.org/10.1111/j.1365-2664.2008.01576.x

Chapter 3

Mechatronics and Robotics in Plant Protection

Rathinavel S., A. P. Mohankumar, R. Kavitha, A. Surendrakumar and Siva M.

Tamil Nadu Agricultural University, Coimbatore, India

Abstract

Plant protection is a critical aspect of agriculture, encompassing various suboperations such as weeding, spraying, and pest–disease–rodent control. Recently, the integration of mechatronic and robotic systems in these areas has garnered significant technological interest. These advancements, driven by the industrial revolution and rapid progress in the agricultural sector, have yielded notable results in farm mechanization, with plant protection emerging as a key focus. Innovative approaches in plant protection include robotic weeders, drone-based spraying, autonomous spraying, site-specific actions, and artificial intelligence (AI)-powered machinery. These technologies offer precise efficient solutions for plant protection reducing the reliance on manual labor and higher chemical inputs. Power system, navigation system, data acquisition and processing, and application systems of plant protection robots were discussed. Machine learning (ML), mapping, AI, and navigation algorithms were addressed. Developed and/or commercial plant protection robots and mechatronic machines were discussed. Sensors detect and monitor environmental conditions, while actuators execute specific tasks based on the data collected. Despite the promising potential of mechatronic and robotic systems in plant protection, several technical and operational challenges that exist are also listed. Sustained research and development in this field are essential to overcome

Sustainable and Smart Farming Eco-systems:
Changing Landscapes in the Age of Technology, 31–45

Published under exclusive licence by Emerald Publishing Limited
doi:10.1108/978-1-83608-092-320261003

existing challenges and unlock the full potential of mechatronics and robotics in plant protection.

Keywords: Precision agriculture; crop protection; site-specific spraying; variable rate technology; artificial intelligence

Introduction

The evolution of agriculture, which began approximately 12,000 years ago during the Neolithic period, marked a significant shift in human civilization. The transition from a hunter-gatherer lifestyle to settled farming communities brought about a dependable food supply and laid the foundation for modern agriculture. Over time, industrial agriculture, characterized by large-scale monoculture, became the dominant force in food production. Despite this, nearly 2 billion people continue to rely on subsistence agriculture.

Looking ahead to 2050, the United Nations projects the global population will reach 9.7 billion (Rathinavel et al., 2023), requiring a 70% increase in agricultural output to meet the demand (Beltran-Peña et al., 2020). This challenge is compounded by the need to achieve the 2030 Agenda's goal of eradicating hunger and ensuring year-round access to nutritious food for all, particularly the most vulnerable populations. Furthermore, the food industry is a major contributor to global energy consumption and greenhouse gas emissions, underscoring the necessity of not only increasing food production but also doing so sustainably.

In response to these challenges, smart farming has emerged as a pivotal innovation (Fig. 3.1). Leveraging technologies from the Fourth Industrial Revolution, smart farming seeks to enhance agricultural productivity and sustainability by optimizing resource use and minimizing environmental impact. This approach integrates advanced data acquisition and analysis to manage farm operations effectively. Unlike traditional technologies, smart farming systems utilize data from various sources – historical, geographical, and instrumental – to provide actionable insights. By employing hardware (Internet of Things (IoT)) and software, smart farming facilitates comprehensive management of agricultural activities, both pre- and post-harvest, ensuring that data are organized, accessible, and actionable from anywhere in the world.

The integration of mechatronics and robotics in plant protection (An et al., 2022) represents a critical advancement within smart farming, offering innovative solutions to safeguard crops and enhance overall agricultural efficiency.

Plant Protection

Plant protection is the method or practice of protecting the plant yields from different agents of pests including weeds, insects, plant diseases, small animals, birds, and other organisms that cause damage to the plants.

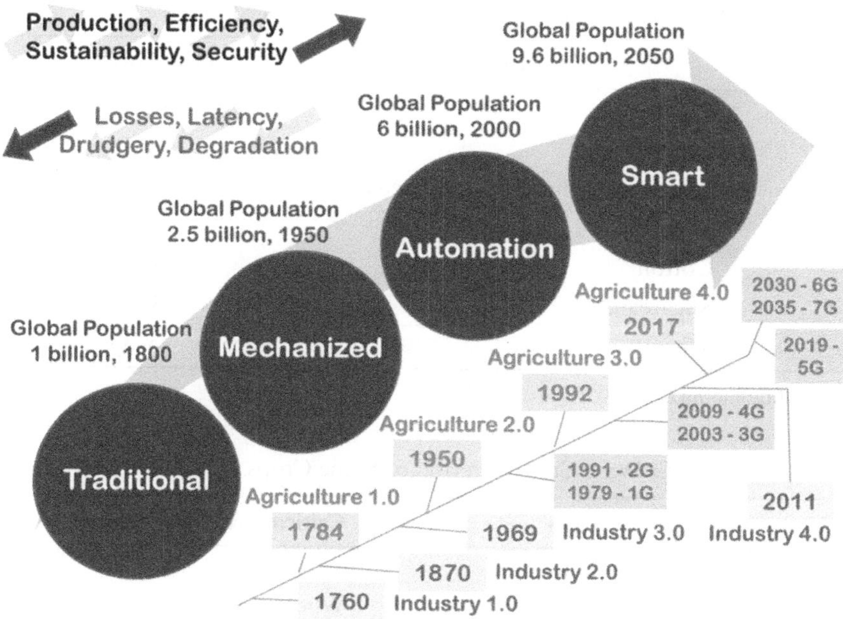

Fig. 3.1. Dynamics of Population, Agriculture Technology, and Industrial Growth. *Source*: Rathinavel et al. (2023).

Weeding

Weeding refers to the process of removing unwanted or invasive plants (weeds) from gardens, farms, orchards, or cultivated areas. It involves the manual, chemical, or mechanical removal of these undesirable plants to ensure the healthy growth of crops.

Around 30,000 weed species were found globally, in which 18,000 of them can cause damage to the plants and about 250 were said to be prominent (Marimuthu et al., 2022). Weeds impact not only crops through potential yield loss (Table 3.1) to plants but also humans and animals, serve as a host for pests, and cause wear and excessive wear and tear of machines like harvesters. The occurrence of weed may be scattered, barren, inter-row, and intra-row.

Plants are very responsive to weed competition and need to be kept free from weeds, from crop establishment, emergence, or bud break, until the end of their critical weed control period. Critical period of weed control (Table 3.2) has been defined as an interval in the life cycle of the crop when it must be kept weed-free to check the yield loss. In mechanization perspective, the critical weed control period is the notable period for planning, management, and mobilization of weeding machinery.

Various weeding methods (Fig. 3.2) such as manual, mechanical, spraying, soil solarization, high heat/burning, etc. were followed in agricultural fields.

Table 3.1. Yield Loss on Some Major Crops due to Weeds.

S. No.	Crop	Yield Loss (%)
1	Rice	30–35
2	Onion	90.7
3	Maize, sorghum, pulses, oilseeds	18–85
4	Sugar beet	48.4
5	Cotton	47.5
6	Sugarcane	38.8
7	Wheat	15–30

Source: Marimuthu et al. (2022).

Table 3.2. Critical Weed Control Period of Some Crops.

S. No.	Crop	Critical Weed-free Period
1.	Transplanted rice	15–45 days
2.	Upland rice	Whole season
3.	Wheat	30–45 days
4.	Cotton	15–60 days
5.	Onions	Whole season
6.	Potatoes	4 weeks after planting
7.	Sugarcane	30–120 days
8.	Groundnut	30–50 days
9.	Soybean	15–45 days
10.	Sorghum	15–45 days
11.	Maize	15–45 days
12.	Carrot	3–6 weeks after emergence
13.	Beets	2–4 weeks after emergence
14.	Lettuce	3 weeks after planting
15.	Sunflower	30–45 days
16.	Garlic	30–70 days after sowing
17.	Cucumbers	4 weeks after seeding
18.	Cabbage	3 weeks after planting
19.	Cauliflower	30–45 after sowing
20.	Sesame	15–45 days

Table 3.2. (*Continued*)

S. No.	Crop	Critical Weed-free Period
21.	Strawberries	During May and June
22.	Brinjal	20–60 after sowing
23.	Tomato	20–60 after sowing
24.	Castor	30–60 after sowing
25.	Finger millet	25–45 days

Source: Tamil Nadu Agricultural University (TNAU) agritech website.

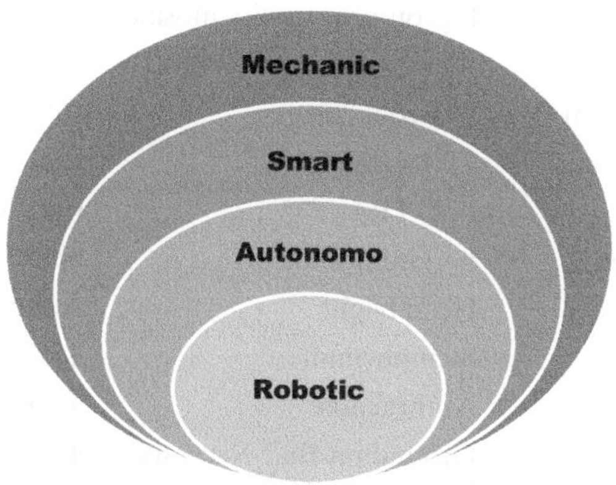

Fig. 3.2. Mechanical Weeding Technologies. *Source*: Authors' compilation.

Insect Pests and Diseases

Insect pests and diseases cause unexpected losses in various crops. Also, nowadays, the climate change is also crucial to understand the behavioral change on pests and reaction of crops. According to the International Rice Research Institute, over 100 species of insects act as pests to rice crops around the world. And out of the 100, about 20 species can cause economic damage in rice. Leaf miner pests in tomatoes can cause 80%–100% losses (Sanda et al., 2018). Flea beetles (*Podagrica* spp.), cotton leafworm cotton stainer bug (*Dysdercus* sp.), egg fruit borers (*Daraba laisalis* Walker), medfly (*Ceratitis capitata*), whitefly (*B. tabaci Gennadius*), etc. are some of the insect pests (Ofuya et al., 2023). Some of the notable pests and diseases of various crops are listed in Table 3.3.

Plant Protection Robots

Plant protection robots represent a transformative advancement in modern agriculture, integrating robotics, mechatronics, and automation to enhance crop management and health. These sophisticated systems leverage precision technologies,

Table 3.3. List of Notable Pests and Diseases Internationally.

S. No.	Plant	Disease	Insect
1.	Rice	Bacterial blight	Water weevil
2.	Maize	Maize lethal necrosis	Fall armyworm
3.	Wheat	Wheat blast	Stem borer
4.	Sugarcane	Pokkah boeng	Stem borers
5.	Coconut	Root wilt	Rhinoceros beetle
6.	Coffee	Coffee wilt disease	Coffee berry borer
7.	Soyabean	Soybean rust	*Helicoverpa armigera*
8.	Red gram	Pigeon pea sterility mosaic	*Helicoverpa armigera*
9.	Bengal gram	Ascochyta blight	*Helicoverpa armigera*
10.	Black gram	Dry root rot	*Helicoverpa armigera*
11.	Ground nut	Tikka leaf spot	Leaf minor
12.	Banana	FoC TR4	Pseudo stem weevil
13.	Sorghum	Sorghum smut	Shoot fly
14.	Sunflower	Head blight	Avian pests
15.	Onion	Stemphylium	Thrips
16.	Capsicum	Anthracnose	Thrips
17.	Tomato	Groundnut Bud Necrosis Virus (GBNV)	Tomato pinworm
18.	Mango	Mango malformation	Mango stem borer
19.	Grapes	Downy mildew	Stem gridler
20.	Cauliflower	Black leg	Dimond back moth

Source: Authors' compilation.

including drones and automated machinery, to perform targeted tasks such as weeding, spraying, and monitoring. By utilizing advanced sensors and AI, plant protection robots can accurately identify and address pest infestations, disease outbreaks, and nutrient deficiencies, thereby reducing the need for broad-spectrum chemical applications. This precision not only improves crop yield and quality but also promotes sustainable farming practices by minimizing environmental impact and optimizing resource use.

Major hardware components of plant protection robotics were as follows:

1. Power system
2. Navigation systems
3. Data acquisition system
4. Control system
5. Application system.

Power System

A power system for robots typically contains a set of batteries or a combination of batteries or solar panels (Averill et al., 2022; Das et al., 2022; Thakare et al., 2024). Solar panels may provide power to continuously operate the robots in the agricultural field, except orchard fields. Voltage regulators are used to supply necessary power input to the controllers and sensors.

Navigation System

Any farm robot is designed to autonomously guide itself through the crop fields or orchards, exactly identifying the targets (weeds, infected plants, pests, etc.,) while avoiding healthy crops. Navigation system (Kanagasingham et al., 2020) utilizes global positioning system (GPS), sensors like encoders, proximity detectors, depth cameras for precise path planning, obstacle detection, and real-time adjustments, for efficient and effective robotic operation. Some of the algorithms used for robotic navigation system developments are as follows:

- Simultaneous localization and mapping (Lajoie et al., 2021).
- Potential fields algorithm (Miao et al., 2021).
- Ordinary least square method (He et al., 2021).
- Extended Kalman filter (Kotaniemi et al., 2024).
- Dijkstra's algorithm (Zhang et al., 2021).
- Monte Carlo localization (Ding et al., 2022).

Galileo, Global Navigation Satellite System (GLONASS), Global Navigation Satellite System (GNSS), and Navstar GPS are some of the GPSs used in different parts of the globe for positioning. For centimeter-level accurate positioning, Real-time Kinematics (RTK) GPS is used, which is essential for working in closely spaced crops and weeds. Inertial Measurement Units are crucial for maintaining stability and orientation, especially on hilly and uneven terrain.

Data Acquisition System

This system collects the information on plants for its existence/non-existence, health status, classification between crop and weed. The system can be classified as a real-time system and an offline system. In the real-time system, it uses sensors like depth camera, GreenSeeker, Yara N-sensor, etc. mounted in robots to extract data on real-time – on-the-go acquisition. Offline data acquisition includes mapping the field data followed by data processing using suitable software to identify the targets with spatial information.

Real-time target detection methods for a plant protection robot can be done via ML techniques. This involves several steps (Liakos et al., 2018; Liu & Bruch, 2020) including data acquisition, pre-processing, feature extraction, and classification.

Map-based plant protection robots get helps from a supplementary system running prior to the main robot's operation. Supplementary systems may be a field robot or a drone with spectral sensors. Spectral sensors may be multispectral or hyperspectral in nature. The hyperspectral vision-based classification of multi-season weed mapping was discussed by Zhang et al. (2012).

Control System

Control systems may have any of the components as master control hardware like microcontroller, microprocessor, flight controller, etc. Microcontrollers, including Arduino UNO, ESP 32 (built-in Wi-Fi and Bluetooth), collect the data from single/multiple sensors and control the output devices such as motors (which actuate the spray or mechanical tool). Processors like Raspberry Pi 4 and NVIDIA Jetson Nano can work for advanced machines embedded with machine vision systems requiring image processing and similar AI applications.

Some of the mechatronic and robotic works on plant protection through ML and sensor-based technology were listed in Table 3.4.

Table 3.4. Various Mechatronic and Robot Developments for Plant Protection Based on Microcontrollers.

S. No.	Development	Description	Reference
1.	Intra-row weeding (tractor connectable)	Uses ultrasonic sensor, fuzzy logic, four-bar linkage mechanism, microcontroller circuit	Kumar et al. (2024)
2.	Intelligent weeding robot	Deep learning model, laser beam, mobile application, Bluetooth module, Arduino microcontroller	Zatari et al. (2022)
3.	Autonomous pesticide spraying robot	Leaf color sensor (TCS 3200), microcontroller board (Micro-Arduino)	Tahmasebi et al. (2022)
4.	Crop and weed classification	ML algorithm, Raspberry Pi	Joshitha et al. (2022)
5.	Smart pesticide spraying robot	ML, Raspberry Pi 3, L293d motor driver, Python	Chaitanya et al. (2020)
6.	Autonomous pesticide spraying robot	ML, Raspberry Pi 3, support vector machine	Kameswari et al. (2022)

Source: Authors' compilation.

Software and Algorithms

Computer programming such as "C" language, C++ language, and Python through development boards like Arduino, ESP 32, Raspberry Pi, etc. can operate the robots.

Various learning models in ML (Liakos et al., 2018) involve

- Regression
- Clustering
- Bayesian models
- Instance-based models
- Decision trees
- Artificial neural networks
- Support vector machines
- Ensemble learning
- You only look once (YOLO).

A robot operating system is a flexible framework for writing robot software, and it is commonly useful for developing navigation and control systems for robots and autonomous machinery.

Software like LabVIEW are in usage for integrating and processing data from sensors like cameras, Light Detection and Ranging (LIDAR), multispectral cameras, etc.

Mapping software like Pix4D mapper, AgiSoft, ArcMap, QGIS, etc., assist in data processing and map generation along with spatial data which serve as an input data for plant protection drones, robots, and mechatronic machinery.

Developed/Commercially Available Robots

Several research institutions and companies are actively developing and refining weeding robots to make weeding more effective and accessible for farm use. Aiming to address challenges in weed management and promoting sustainable agricultural production, such robots were invented, and some of them were discussed as follows:

Dick, Tom, and Harry: It is an autonomous nonchemical weeding system. It is the world's first nonchemical robotic weeding for arable crops. A small robot company, a British agritech startup for sustainable farming, has developed AI-enabled robots – named Tom, Dick, and Harry – that can identify and kill individual weeds with electricity. These agricultural robots could reduce the use of harmful chemicals and heavy machinery, paving the way for a new approach to sustainable crop farming. It will integrate Rootwave's weed zapping technology to make autonomous nonchemical weeding a reality. It's the best of all worlds – having an increased yield as well as minimal chemical usage through being more precise. Tom – It is a crop monitoring and mapping/scanning robot that lives on the farm covering 20 ha per day, using "Wilma," an AI operating system. These data can then be sent to Dick – a nonchemical robotic weeding system that zaps individual weeds with electrical "lightning strikes." And finally, Harry plants seeds in the weed-free soil.

See & Spray Technology: John Deere, the 180-year-old manufacturer of farming and industrial machinery, is developing advanced equipment designed to reduce herbicide usage by 80%–90% through vision processing and ML technology. "See & Spray" is a groundbreaking approach that uses computer vision and ML to spray herbicides precisely where weeds are present. The key innovation behind See & Spray lies in the embedded Nvidia Jetson Xavier, a graphics processor trained to distinguish between weeds and crop plants. Two models have been developed: See & Spray Select and See & Spray Ultimate. See & Spray Select helps farmers reduce non-residual, pre-emergent herbicide use by 77% on average, targeting weeds in fallow ground. See & Spray Ultimate offers "green-on-green" in-season targeted spraying for fields planted with corn, cotton, and soybeans in 30-inch or wider rows, as well as broadcast spraying capabilities.

Duck: An engineer at Japanese automaker Nissan has developed a robot duck designed to help farmers reduce herbicide and pesticide use in rice fields. This compact robot mimics the natural behavior of ducks, which swim in flooded paddy fields, uproot weeds, consume insects, and fertilize the soil with their manure. Named "Aigamo," after the breed of duck used in this traditional method, the robot weighs 1.5 kg and is roughly the size of a large robotic vacuum. Its underside features two rotating rubber brushes that mimic a duck's feet, oxygenating the water and preventing weed growth. The robot operates using Wi-Fi, batteries, solar power, and GPS to navigate the fields.

Bakus: VitiBot is a French industrial company, on the market of the electric vineyard straddles with the vocation to work in full autonomy. VitiBot (Milburn et al., 2023) developed Bakus, an unmanned automatic orchard robot. It is equipped with 360° infrared vision and weight balancing. Bakus is a full electric and autonomous vineyard robot. It is capable of carrying out most of the viticultural work, nowadays carried out by conventional tractors, in the best conditions of safety and security. Bakus integrates a wide range of modular tools for an efficient soil work in vineyards. The tool holders can be easily integrated on either side of the Bakus structure and allow very precise adjustment of chosen working height according to the tools. One can use electric intervines, electric crush mower, simple disc harrow, kress fingers, agromy pack, full mechanical weeding tools, lump-breaker discs, and many others. Used for a full mechanical or electrical soil work and soon vine works.

IBEX: IBEX Automation Ltd develops autonomous agricultural robot systems for farmers, including an autonomous precision weed detection and spraying system for use in grassland. IBEX (Raj et al., 2020) is one meter long and is a fully autonomous all-terrain vehicle. Also, the caterpillar wheels help IBEX to conquer the tough territories and slopes of up to 45 degrees. IBEX is primarily used for destroying weeds. A high-resolution on-board camera and navigation system helps in detection of weeds. Images from the camera are utilized to observe the condition of the plants and soil. Furthermore, a manual control is also possible due to the live streaming of the camera feed. This feature allows farmers to make suitable changes in a specific area whenever necessary.

Oz, Dino, and Ted: These Oz, Dino, and Ted are mechanical robotic weeders (Bangale & Kumar, 2024). Developed by Naïo Technologies, this develops and markets robots for agriculture and viticulture. Dino navigates the field autonomously with a 2-cm precision range with a guidance system that combines the information from RTK GPS and other sensors. Dino detects crop rows and adjusts the tools to weed as close to the plants as possible. Ted is a vineyard which involves effective and precise mechanical weeding, without herbicides, which respects our soil and crops. Ted is easy to use. The universal mounting frame offers the possibility to adapt various tools. Oz also works completely autonomously with its RTK GPS guidance system. Oz helps hoeing, weeding, making furrows, seeding, transport, and assisting farmers.

Crop Weed Signaling: A novel and very effective technique (Raja et al., 2019; Su et al., 2020) for robotic real-time weed control is crop detection using exogenous fluorescent signals. Those signals are to be generated from crop plants instead of weeds through application or enhancement of suitable markers (Nguyen et al., 2017). Fluorescence markers, currently researched, are biological markers, physical markers, and chemical markers for crop weed signaling techniques.

The concept of plant labels is to have an inexpensive and biodegradable label material such as straw, attached with the crop plants to have fluorescent signaling properties. This is done by painting a component with fluorescent paint and attached with saplings. The saplings are transplanted with the painted label material into the field which exhibits inflorescence on exposure to ultraviolet (UV) light. Another method is to apply the signaling compounds directly into a part of the seedling (stem/foliage). When the robotic weeding machine comes across the plant with any of the above discussed signaling compounds (marker), the UV light is passed under a closed condition to get back a fluorescent signal which aids in recognizing the crop plant. Further, the machine acts around the crop without causing any damage to the crop. These signaling compounds (markers) must be unique, safe, have sufficient signal strength and lifetime, leakage free, stable, not affect crop growth, and also cost-effective. Systemic marker involves chemical treatment of roots with suitable chemical liquids, which leads crops to have fluorescent signals on their leaves when exposed to UV light. Tomato and lettuce crops (Su, 2020) were experimented with the crop signaling technique and successful at laboratory levels.

Variable Rate Spraying System: Applies pesticides, herbicides, or fertilizers at varying rates across a field based on specific needs. The system may use any of the detection technologies discussed earlier and adopts variable flow control architecture and hardware for achieving variable rate spraying systems at different rates of inputs across the same field based on the severity/demands. Table 3.5 discusses the various research works carried out on variable rate spray systems.

Agrobot Bug Vacuum Machine: With two big fans, this machine is driving in pattern to clean the bugs from your field. The Bug Vacuum machine has a double-fan vacuum system to ensure a uniform airflow over the whole bed width. An optimum vacuum pressure is achieved due to its precise fan-height control, maximizing bug aspiration. In compliance with food safety standards, the fan inlet is made of stainless steel and food-grade silicone. Power is 20 hp, vacuum power is 2×4 hp, and wind speed is 35 miles/h.

Table 3.5. Developments in Variable Rate Spray Systems for Plant Protection.

S. No.	Development	Description	Reference
1.	Precision variable rate spray system for Unmanned Aerial Vehicles (UAV)	Single-chip microcomputer and micro diaphragm pump	Lian et al. (2019)
2.	Independent aerial sprayer system for precision agriculture with vision sensors	Red Green Blue (RGB) camera and small computer, visual-based RGB Vegetation Indices assessment	Agurob et al. (2023)
3.	Smart variable rate sprayer	Deep learning using YOLOv3 and tiny-YOLOv3 models	Hussain et al. (2020)
4.	Site-specific pesticide application	Site-specific applications compared to conventional applications	Rajmis et al. (2022)
5.	Sensor-based site-specific sprayer	RGB color sensor (TCS3200), Green seeker device and spectral sensor (AS7341, 10-channel)	Anand (2020)

Source: Authors' compilation.

Conclusion

Robots that protect plants have a number of benefits to contemporary agriculture that can enhance efficiency, sustainability, and productivity. Among the advantages, precision and accuracy should be listed as one of the most important ones as they increase the effectiveness of the production through the opportunity to make pesticides or fertilizer or other treatments targeted. The site-specific management is also supported by these robots, which enable farmers to utilize resources in the best way depending on the specific requirements of certain areas of a field. The second major benefit is that it reduces labor dependence, which assists in countering labor shortages, and to a large extent, it will reduce physical labor that is characterized by traditional farming practices. Moreover, a large number of plant protection robots are powered either by solar energy or by electric tools, thus being more friendly to the environment as they do not rely on fossil-based energy-powered equipment. They also guarantee harmonization of operations which may result in quality produce and improved grading. In addition, these robotic systems could work 24 hours in the day, which would accelerate productivity and enable farm operations even in the event of labor limitations.

Along with these advantages, there are a number of concerns related to the application of AI and robotics in the agricultural industry as well. Leslie (2019) and Weidinger et al. (2021) emphasize that AI systems are not always clear, understandable, and reliable, and it is challenging to support or analyze their conclusions. It is also probable that AI-based systems might exhibit bias or discriminatory tendencies as a result of the nature of data used to train them. Such technologies can also result in the deprivation of specific rights and a decreased level of autonomy on the personal level and a limited number of appeals in the case of wrong decisions of automated systems. More to the point, the gathering and analysis of significant volumes of agricultural and individual data may cause privacy issues. Other researchers also suggest that further automation can lead to social disintegration and isolation particularly in rural areas where traditional labor relations are a social factor. Lastly, malicious or unethical application of AI technologies may also be present, which may cause additional risks in the future and lead to their misuse or manipulation.

To summarize, the plant protection robots are an important development in contemporary farming since they enhance precision, efficiency, and sustainability in agricultural activities. They can significantly increase the productivity of agriculture and optimize its resources since they can carry out site-specific management, work around the clock, and minimize the use of manual labor. Nevertheless, there are also significant ethical, social, and technological issues like transparency, privacy, bias, misuse, etc. linked with AI and robotics integration. Thus, as these technologies provide good prospects in the future of agriculture, they need to be implemented with proper regulations, ethics, and protection measures to achieve responsible and fair use.

References

Agurob, M. C., Bano, A. J., Paradela, I., Clar, S., Aleluya, E. R., & Salaan, C. J. (2023). Autonomous Vision-based unmanned aerial spray system with variable flow for agricultural application. *International Journal of Computer Science, 50*, 3.

Anand, R. (2020). Development of sensor-based site-specific sprayer for disease control in vegetable crops [Unpublished doctoral thesis]. Division of Agricultural Engineering ICAR – Indian Agricultural Research Institute, New Delhi, India.

An, Z., Wang, C., Raj, B., Eswaran, S., Raffik, R., Debnath, S., & Rahin, S. A. (2022). Application of new technology of intelligent robot plant protection in ecological agriculture. *Journal of Food Quality*, 1257015.

Averill, K. M., Westbrook, A. S., Pineda-Bermudez, L., O'Briant, R. P., DiTommaso, A., & Ryan, M. R., 2022. Effects of Tertill® weeding robot on weed abundance and diversity. *Agronomy*, *12*(8), 1754.

Bangale, R., & Kumar, M. (2024). Robot-based weed identification and control system. In *Precision agriculture for sustainability* (pp. 169–194). Apple Academic Press.

Beltran-Peña, A., Rosa, L., & D'Odorico, P. (2020). Global food self-sufficiency in the 21st century under sustainable intensification of agriculture. *Environmental Research Letters*, *15*(9), 095004.

Chaitanya, P., Kotte, D., Srinath, A., & Kalyan, K. B. (2020). Development of smart pesticide spraying robot. *International Journal of Recent Technology and Engineering*, *8*(5), 2193–2202.

Das, G. P., Gould, I., Zarafshan, P., Heselden, J., Badiee, A., Wright, I., & Pearson, S. (2022). Applications of robotic and solar energy in precision agriculture and smart farming. In *Solar energy advancements in agriculture and food production systems* (pp. 351–390). Academic Press.

Ding, H., Zhang, B., Zhou, J., Yan, Y., Tian, G., & Gu, B. (2022). Recent developments and applications of simultaneous localization and mapping in agriculture. *Journal of Field Robotics*, *39*(6), 956–983.

He, C., Chen, Q., Miao, Z., Li, N., & Sun, T. (2021). Extracting the navigation path of an agricultural plant protection robot based on machine vision. In *IEEE 40th Chinese control conference (CCC)* (pp. 3576–3581). IEEE.

Hussain N., Farooque A. A., Schumann A. W., McKenzie-Gopsill, A., Esau T., Abbas F., Acharya, B., & Qamar Zaman, Q. (2020). Design and development of a smart variable rate sprayer using deep learning. *Remote Sensing*, *12*, 4091.

Joshitha, K. L., Nishanthini, S., Harshini, B. J., & Sadhana, T. (2022). Raspberry Pi based crop and weed classification using machine learning algorithm. *IOP Conference Series: Earth and Environmental Science*, *980*(1), 012029.

Kameswari, S. S. D., Prabhakar, T., & Kishore, K. K. (2022, January). Autonomous pesticide spraying robot using SVM. In *Proceedings of international conference on wireless communication: ICWiCom* (pp. 293–302). Springer Nature.

Kanagasingham, S., Ekpanyapong, M., & Chaihan, R. (2020). Integrating machine vision-based row guidance with GPS and compass-based routing to achieve autonomous navigation for a rice field weeding robot. *Precision Agriculture*, *21*(4), 831–855.

Kotaniemi, J., Känsäkoski, N., & Heikkilä, T. (2024). *A weeding robot for seedling removal.* arXiv:2405.12596.

Kumar, S. P., Tewari, V. K., Chandel, A. K., Mehta, C. R., Nare, B., Chethan, C. R., & Mundhada, K. (2024). Automated prototype intra row weeding system. *Journal of Scientific & Industrial Research*, *83*, 362–374.

Lajoie, P. Y., Ramtoula, B., Wu, F., & Beltrame, G. (2021). *Towards collaborative simultaneous localization and mapping: A survey of the current research landscape*. arXiv preprint arXiv:2108.08325.

Leslie, D. (2019). *Understanding artificial intelligence ethics and safety*. arXiv preprint arXiv:1906.05684.

Liakos, K. G., Busato, P., Moshou, D., Pearson, S., & Bochtis, D. (2018). Machine learning in agriculture: A review. *Sensors*, *18*(8), 2674.

Lian, Q., Tan, F., Fu, X. M., Zhang, P., Liu, X., & Zhang, W. (2019). Design of precision variable-rate spray system for unmanned aerial vehicle using automatic control method. *International Journal of Agricultural and Biological Engineering*, *12*(2), 29–35.

Liu, B., & Bruch, R. (2020). *Weed detection for selective spraying: A review*. Current Robot Repository.

Marimuthu, S., Pavithran, P., & Gowtham, G. (2022). *Polymeric systems for the delivery of herbicides to improve weed control efficiency*. IntechOpen.

Miao, Z., Yang, G., He, C., Li, N., & Sun, T. (2021, October 22–24). Artificial potential field method for area coverage of multi agricultural robots. In *Intelligent equipment, robots, and vehicles: Proceedings 7th international conference on life system modeling and simulation, LSMS 2021 and 7th international conference on intelligent computing for sustainable energy and environment, ICSEE 2021, Hangzhou, China* (Part III, 7, pp. 67–76). Springer.

Milburn, L., Gamba, J., Fernandes, M., & Semini, C. (2023). Computer-vision based real time waypoint generation for autonomous vineyard navigation with quadruped robots. In *IEEE international conference on autonomous robot systems and competitions (ICARSC)* (pp. 239–244). IEEE.

Nguyen, T. T., Slaughter, D. C., Fennimore, S. A., & Vuong, V. L. (2017). Designing and evaluating the use of crop signaling markers for fully automated and robust weed control technology. In *ASABE annual international meeting* (Paper No. 1700160, pp. 1–5). ASABE.

Ofuya, T. I., Okunlola, A. I., & Mbata, G. N. (2023). A review of insect pest management in vegetable crop production in Nigeria. *Insects*, *14*(2), 111.

Raj, S., Aatitya, R. M., Samuel, S. J., Karthik, J. V., & Ezhilarasi, D. (2020). Ibex: A reconfigurable ground vehicle with adaptive terrain navigation capability. In *2020 IEEE international conference on robotics and automation (ICRA)* (pp. 2975–2980). IEEE.

Raja, R., Slaughter, D. C., Fennimore, S. A., Nguyen, T. T., Vuong, V. L., Sinha, N., Tourte, L., Smith, R. F., & Siemens, M. C. (2019). Crop signaling: A novel crop recognition technique for robotic weed control. *Biosystems Engineering*, *187*, 278–291.

Rajmis, S., Karpinski, I., Pohl, J., Herrmann M., & Kehlenbeck, H. (2022). Economic potential of site-specific pesticide application scenarios with direct injection and automatic application assistant in northern Germany. *Precision Agriculture*, *23*, 2063–2088.

Rathinavel, S., Kavitha, R., Gitanjali, J., & Saiprasanth, R. (2023). Role of 5G technology in enhancing agricultural mechanization. *IOP Conference Series: Earth and Environmental Science*, *1258*(1), 012010.

Sanda, N. B., Sunusi, M., Hamisu, H. S., Wudil, B. S., Sue, H., & Abdulahi, A. M. (2018). Biological invasion of tomato leaf miner, Tuta absoluta (Meyrick) in Nigeria: Problems and management strategies opitimization: A review. *Asian Journal of Agricultural and Horticultural Research*, *1*, 1–14.

Su, W. H. (2020). Systemic crop signaling for automatic recognition of transplanted lettuce and tomato under different levels of sunlight for early season weed control. *Challenges*, *11*(2), 23.

Su, W. H., Fennimore, S. A., & Slaughter, D. C. (2020). Development of a systemic crop signaling system for automated real-time plant care in vegetable crops. *Biosystems Engineering*, *193*, 62–74.

Tahmasebi, M., Gohari, M., & Emami, A. (2022). An autonomous pesticide sprayer robot with a color-based vision system. *International Journal of Robotics and Control Systems*, *2*(1), 115–123.

Thakare, V., Sunheriya, N., Khedkar, S., Kamble, P., & Sahare, S. (2024, August). Application of solar energy with weeding robot in agriculture field. *AIP Conference Proceedings*, *3139*(1), 030017.

Weidinger, L., Mellor, J., Rauh, M., Griffin, C., Uesato, J., Huang, P. S., & Gabriel, I. (2021). *Ethical and social risks of harm from language models*. arXiv preprint arXiv:2112.04359.

Zatari, A., Hoor, B., & Nasereddin, N. (2022). Intelligent weeding robot using deep-learning [Undergraduate thesis, Palestine Polytechnic University, Palestine].

Zhang, L., Li, R., Li, Z., Meng, Y., Liang, J., Fu, L., Jin, X., & Li, S. (2021). A quadratic traversal algorithm of shortest weeding path planning for agricultural mobile robots in cornfield. *Journal of Robotics*, *1*, 6633139.

Zhang, Y., Slaughter, D. C., & Staab, E. S. (2012). Robust hyperspectral vision-based classification for multi-season weed mapping. *ISPRS Journal of Photogrammetry and Remote Sensing*, *69*, 65–73.

Chapter 4

Intelligent Agriculture and Sustainable Food Systems in India

Sanitha A. C.

School of Business and Economics Studies, Chinmaya Vishwa Vidyapeeth, Deemed to be University, Ernakulam, Kerala, India

Abstract

India's agriculture sector, a cornerstone of its economy, supports millions of livelihoods and significantly contributes to national income. Yet, this sector faces formidable challenges, including water scarcity, soil degradation, and the impacts of climate change. Addressing these issues is crucial for ensuring food security and fostering sustainable economic growth. In this context, smart agriculture, which integrates advanced technologies such as precision farming and sensor-based irrigation systems, offers promising solutions. These technologies aim to enhance agricultural productivity, optimize resource use, and promote environmental sustainability. This chapter explores the current landscape of smart agriculture and sustainable food systems in India. It examines the adoption and impact of these technologies across various agro-climatic zones and states, assessing their effects on productivity, resource efficiency, and sustainability. Additionally, the study evaluates existing policy frameworks and institutional support, identifying opportunities and challenges in scaling successful practices. The research employs a qualitative methodology, utilizing secondary data sources such as government reports, academic publications, and statistical data. Key sources include reports from the Ministry of Agriculture and Farmers' Welfare, research articles from journals like the Journal of Agricultural Technology, and data from the National Statistical Office. In conclusion, smart agriculture and sustainable food systems represent transformative solutions to the challenges facing Indian agriculture. By integrating advanced technologies with sustainable practices, India can enhance productivity, conserve resources, and increase resilience to climate

Sustainable and Smart Farming Eco-systems:
Changing Landscapes in the Age of Technology, 47–57

doi:10.1108/978-1-83608-092-320261004

change. Effective policy frameworks, investment in research, and collaborative efforts are essential for scaling these technologies and achieving sustainable agricultural development.

Keywords: Intelligent agriculture; sustainable food systems; agricultural productivity; resource efficiency; environmental sustainability; technology adoption; policy support; food security; climate change; water scarcity; soil health

Introduction

India's agriculture sector is a cornerstone of its economy, providing livelihoods to millions and contributing significantly to national income. However, it faces critical challenges such as water scarcity, soil degradation, and the impacts of climate change. Addressing these challenges is essential for ensuring food security and promoting sustainable economic growth. Smart agriculture, which incorporates advanced technologies such as precision farming and sensor-based irrigation systems, offers promising solutions to these issues. These technologies aim to enhance productivity, optimize resource use, and promote environmental sustainability. The adoption of smart agriculture in India is not uniform, with varying degrees of implementation across different regions and among smallholder farmers. Despite the potential benefits, there are substantial barriers, including high initial costs, lack of awareness, and inadequate infrastructure. Moreover, policy support and investment in research and development play a crucial role in facilitating the transition to smart agriculture. Understanding the regional disparities and socio-economic factors influencing adoption is vital for designing effective interventions. Ultimately, addressing these multi-layered challenges is essential to building climate-resilient architectures for crop management (Saikath & Supriya, 2023). This chapter explores the landscape of smart agriculture and how it intersects with environmentally sustainable food systems to ensure national nutrition security (Arunachalam & Ayyappan, 2013). This chapter explores the landscape of smart agriculture and sustainable food systems in India. It focuses on the adoption and impact of these technologies across various agro-climatic zones and states, evaluating their effects on productivity, resource efficiency, and sustainability. Additionally, the study assesses current policies and support frameworks, identifying opportunities and challenges in scaling successful practices. By analyzing secondary data from government reports, academic publications, and statistical sources, this research aims to provide actionable insights for policymakers, researchers, and stakeholders.

Scope of the Study

This study focuses on smart agriculture technologies and sustainable food systems in India. It examines various agro-climatic zones and states to provide a detailed understanding of the current landscape. The scope includes:

- Adoption rates of technologies such as precision farming and sensor-based irrigation.
- Effectiveness of government initiatives like Pradhan Mantri Krishi Sinchayee Yojana (PMKSY) and National Mission for Sustainable Agriculture (NMSA).
- Challenges and opportunities related to scaling these practices across different regions.

Review of Literature

Numerous studies have explored various facets of smart agriculture and sustainable food systems in India. The exploration of smart agriculture and sustainable food systems in India has expanded to include an emphasis on technological integration, regional equity, and long-term sustainability. Kapoor and Menon (2023) discuss the significant strides India has made at the G20 level, advancing the integration of climate-smart agriculture practices to ensure both food security and environmental sustainability. This is echoed by Ahmad (2023), who focuses on the critical role of Internet of Things (IoT)-driven solutions in facilitating data-driven decision-making and addressing India's complex agricultural challenges. These emerging technologies, which blend data science and traditional practices, are seen as transformative in terms of enhancing efficiency and productivity. Arpula et al. (2024) delve into the specific application of smart irrigation systems, demonstrating how modern technologies can optimize water use in water-scarce regions. Kapoor and Pal (2024) evaluate the economic benefits of these technologies in semi-arid regions, showcasing improvements in farmer incomes due to the adoption of climate-smart practices. Barooah et al. (2023) further explore the relationship between gender and agricultural policies, arguing that gender-inclusive agricultural technologies are crucial to ensuring broad-based benefits for all farmers, particularly women in rural areas. These studies stress the importance of both technological solutions and supportive policies in driving sustainable agriculture. The increasing reliance on organic farming as a sustainable agricultural model is another significant trend. Beulah et al. (2023) highlight the growing post-pandemic shift toward organic and regenerative farming practices, which are seen as vital for long-term environmental health. Tanti et al. (2024) provide a detailed account of how climate-smart agricultural practices not only improve farm income but also contribute to ecological resilience in Eastern India, where the agriculture sector is particularly vulnerable to the effects of climate change. Further, Kumar et al. (2024) discuss the widespread adoption of IoT technologies in agriculture, underlining their potential to foster smart and sustainable farming practices across different regions of India. Meanwhile, Dhanaraju et al. (2022) explore the benefits of IoT-based solutions in providing real-time data for farmers, enhancing decision-making processes. The work of Balkrishna et al. (2023) highlights India's growing digital agricultural infrastructure and the role of artificial intelligence (AI) and big data in optimizing farming practices, while Ranjan et al. (2024) showcase the promise of precision agriculture in maximizing yield efficiency and resource management.

Research Gap

Despite the extensive body of work on smart agriculture and sustainability in India, critical gaps remain that warrant further investigation. One of the primary gaps is the integration of various technological solutions across different farming systems. While many studies focus on the benefits of specific technologies – such as smart irrigation, IoT-based monitoring, and precision farming – there is limited research on how these technologies can be synergistically combined to optimize agricultural practices across diverse agro-climatic zones (Kumar et al., 2024; Arpula et al., 2024). Another important gap lies in the regional disparities in technology adoption. As noted by Saikath and Supriya (2023) the uptake of smart farming technologies is uneven across India, with significant differences between urban and rural areas and between large-scale and smallholder farmers. Factors such as farmer education, financial accessibility, and infrastructure limitations play a major role in these disparities. More research is needed to understand the socio-economic barriers to adoption and how policies can be tailored to address these challenges. A further research gap is the long-term sustainability and scalability of these technologies. While the short-term benefits – such as increased crop yield and resource efficiency – are well documented, fewer studies explore the long-term impact of smart agriculture on environmental sustainability and the resilience of farming systems in the face of climate change (Beulah et al., 2023; Dhanaraju et al., 2022). More longitudinal studies are needed to assess the durability of these technologies over time and in varying environmental conditions. Additionally, while government policies, such as the PMKSY and NMSA, have been discussed in the context of promoting sustainable practices, there is limited evaluation of their effectiveness in fostering technological innovation and meeting local needs. Studies are needed to critically assess the alignment of national policies with regional agricultural realities and their success in facilitating the widespread adoption of smart agriculture practices (Kapoor & Menon, 2023). Finally, social equity in the context of smart agriculture is an underexplored area. As highlighted by Barooah et al. (2023), agricultural technologies must be inclusive, particularly for women and marginalized farmers, to ensure equitable benefits. Further research is needed to understand how these technologies can be tailored to address the unique needs of these groups, ensuring that smart agriculture does not exacerbate existing social inequalities. Addressing these gaps will offer a more comprehensive understanding of the dynamics of smart agriculture and its potential to drive sustainable, inclusive, and resilient agricultural systems in India (Kapoor & Pal, 2024; Tanti et al., 2024).

Significance of the Study

This study is significant as it addresses key gaps in the understanding and application of smart agriculture in India. By evaluating the integration of various technologies such as IoT, precision farming, and smart irrigation, the study highlights how these technologies can work together to optimize agricultural practices across different agro-climatic zones (Kumar et al., 2024; Sharma et al., 2024). It also explores regional disparities in technology adoption, identifying socio-economic barriers and offering insights into overcoming these challenges to ensure equitable access to smart farming tools. The research contributes to the evaluation of the long-term sustainability and scalability of smart agriculture technologies, particularly their impact on environmental resilience and climate change adaptation (Beulah et al., 2023; Dhanaraju et al., 2022). Furthermore, the study critically examines the effectiveness of government policies like PMKSY and NMSA, offering recommendations for improvements that better align these initiatives with local agricultural realities (Kapoor & Menon, 2023). Additionally, the study emphasizes the importance of inclusivity in smart agriculture, particularly for marginalized groups such as women and smallholder farmers, ensuring that technological advancements benefit all farmers (Barooah et al., 2023). Overall, the research provides valuable insights to inform policy, investment decisions, and future research, ultimately supporting India's goals for sustainable agriculture, food security, and climate resilience (Sharma et al., 2024; Tanti et al., 2024).

Objectives

1. *Evaluate Technology Impact*: To assess the effects of smart agriculture technologies on agricultural productivity, resource efficiency, and environmental sustainability in India.
2. *Examine Policy Frameworks*: To evaluate the effectiveness of current policies and institutional support mechanisms in promoting sustainable agricultural practices.
3. *Identify Challenges and Opportunities*: To explore barriers to the adoption of smart agriculture and identify opportunities for scaling successful practices across diverse regions of India.

Research Methodology

The research is qualitative in nature, and it primarily depends on secondary data sources in order to analyze the growth and impact of smart agriculture and sustainable food systems in India. This approach will facilitate an in-depth exploration of the topic through a wide range of existing literature, statistical databases, and policy documents. Utilizing several reliable and diverse sources, the study aims to provide a broad understanding of the subject matter, capable of critically analyzing technological developments and their implications for sustainable agriculture.

Government reports, especially those of the Ministry of Agriculture and Farmers' Welfare, form the foundational basis of this research. Annual reports provide a general overview of India's agricultural performance and highlight major programs in operation. Specific program documents, such as those related to the PMKSY and the NMSA, offer detailed insights into objectives, progress, and outcomes. These reports are invaluable for understanding the policy landscape, resource allocation, and the impact of targeted interventions on agricultural practices and sustainability. Besides government data, the study bases its findings on academic publications and statistical databases.

Peer-reviewed journals, including the *Journal of Agricultural Technology* and the *International Journal of Sustainable Agriculture*, provide information on recent technological advancements, such as IoT-based farming solutions and climate-smart practices, contextualized within the Indian agricultural framework. Some of the national statistical sources are adopted in measuring the level of adoption of smart agricultural technologies such as global positioning system (GPS)-enabled tractors, sensor-based irrigation systems, and drones among other agricultural farm technologies. Even the socio-economic effects of such technologies on farmers' productivity and resource use are identified through these statistical sources. Data from international organizations, such as the Food and Agriculture Organization and World Bank, as well as local nongovernmental organizations, allow for a multidimensional analysis of India in comparison with global trends. The comparative analysis ensures a more robust framework to analyze potential as well as challenges in smart agriculture in India.

Data Analysis

The research involved a comprehensive analysis of the adoption rates of smart agricultural technologies, such as GPS-equipped tractors, sensor-based irrigation systems, and other precision tools across different regions to have in-depth understanding of how well these technologies penetrate and are effective in diverse agricultural settings. The analysis considered socio-economic and infrastructural factors that influence adoption, such as access to technology, farmer education, and regional disparities. In addition, the study assessed the effect of these technologies on crucial dimensions – they are productivity, resource efficiency, and environmental sustainability.

Secondary data analyses set a focus on how these innovations lead to reduction in the wastage of resources, improved crop yields, and reduced degradation of the environment – a step ahead toward sustainable development goals (SDGs) including SDG 2 – Zero Hunger, SDG 12 – Responsible Consumption and Production, and SDG 13 – Climate Action. A comprehensive review of existing agricultural policies and support frameworks to address the established effectiveness in the promotion of smart agricultural practices was carried out. The review would bring out gaps and areas needing improvement, thereby presenting recommendations for strengthening policy alignment with sustainability objectives and further advancing progress toward achieving the SDGs through sustainable agriculture.

This research-based study draws data from a range of credible secondary sources and uses an analytical approach focused on adoption, impact, and policy evaluation to examine the potential of smart agriculture to drive sustainable food systems in India. Insights from this research are expected to be theoretically enriching while contributing practically to policymakers, researchers, and practitioners in the field.

Results and Discussion

The study reveals significant progress in the adoption of smart agriculture technologies in India, highlighting their potential to transform the agricultural landscape. Precision farming techniques have been applied to approximately 5% of India's agricultural land, demonstrating a niche yet growing role in enhancing productivity and resource efficiency. The adoption of GPS-enabled tractors has experienced a robust annual growth rate of 15%, reflecting a rising acceptance of these technologies among Indian farmers for optimizing field operations. Sensor-based irrigation systems, particularly prevalent in states like Punjab and Maharashtra, have notably improved water use efficiency and crop yields. Data indicate that about 30% of farmers in these regions now utilize these systems, leading to better water management and enhanced agricultural outputs (Jain & Sharma, 2022).

Government initiatives such as the PMKSY and the National Mission on Sustainable Agriculture (NMSA) have been pivotal in supporting these technological advancements. PMKSY has significantly contributed to improving water use efficiency, while NMSA focuses on enhancing soil health and promoting organic farming practices. These initiatives have reached over 40% of the country's districts, aligning with national goals for sustainable agriculture. Despite these positive developments, the study identifies persistent challenges such as high initial investment costs, limited awareness among smallholder farmers, and uneven policy implementation. Approximately 60% of smallholder farmers cite cost as a major barrier, while 45% report insufficient knowledge about available technologies.

Opportunities for further advancement include leveraging technological innovations such as AI and precision agriculture tools, which could potentially boost productivity by up to 30% and reduce input costs by 15%. Public–private partnerships have shown promise in scaling smart agriculture practices, with a 25% increase in technology adoption in areas supported by these collaborations. Additionally, enhancing policy support and increasing investment in research and development are crucial for overcoming existing barriers and advancing smart agriculture technologies. These efforts are essential for ensuring food security, improving agricultural productivity, and promoting environmental sustainability in India.

Findings

Smart agriculture technologies are increasingly transforming the agricultural landscape of India with innovative, productive, and sustainable solutions, though adoption is still in its infancy. Precision farming techniques currently cover approximately 5% of the agricultural land in India; in fact, their niche yet growing role is reflected in optimization in the application of farm practices. Efforts to expand the reach of precision agriculture are fueled by increasing awareness and the rapid advancements in technology. Take, for instance, the GPS-enabled tractor segment, which was experiencing an adoption growth rate of 15% per annum, as it gained importance in improving the operations as well as reducing laborious processes. On similar lines, sensor-based irrigation systems are becoming increasingly popular, with about 30% of the farmers in agriculturally developed states such as Punjab and Maharashtra using these systems to manage water better and increase crop yield. These trends indicate a positive trajectory, and for these are complemented by focused government efforts under the PMKSY and the National Mission on Sustainable Agriculture (NMSA), that have reportedly reached over 40% of the districts of the country till date in 2024. These focused programs target water conservation, soil health enhancement, and the integration of climate-smart agricultural practices.

The benefits of smart agricultural technologies are starting to reveal themselves increasingly now. These innovations have been reported to have, on average, contributed to an increase in productivity of 20% as well as about a 25% enhancement in resource use efficiency, meaning that they are relatively capable of raising yields within the availability of critical resources. Such changes support SDGs like SDG 2 – Zero Hunger, SDG 12 – Responsible Consumption and Production, and SDG 13 – Climate Action. Still, there are many challenges yet to be addressed that limit widespread use. High initial investment costs continue to present a key challenge, as 60% of smallholder farmers indicate that this is a significant constraint. A further 45% of farmers argue that they have insufficient or no knowledge/experience with these technologies; thus, there is a real need to support extension services and farmer education programs. The lack of uniform application of agricultural policies contributes to this problem, leading to variation across regions. While developed countries have reached up to 75% of the policy benefit, developing regions lag far behind and reach only up to 50%.

With technology advancement, it is seen that many new opportunities lie with such a great improvement. AI-based precision agriculture and IoT-based solutions can help increase productivity by 30% and reduce input costs by 15%. These technologies, however, mainly benefit a few groups while smallholder farmers are largely denied access due to infrastructural and economic constraints. There is an emerging necessity for public–private partnerships as an important conduit to overcoming such barriers. Recent initiatives supported by partners in some regions have increased uptake of advanced technologies by 25%. This is an example of the effectiveness of partnership-supported efforts in bridging gaps in technology access and implementation. Investment in agricultural research and development has increased by 20% over five years, which reflects the nation's efforts toward innovation and dealing with emerging issues in agriculture. Such investment would be central to building cutting-edge solutions that may well make agriculture more sustainable, resilient, and socially inclusive and contribute significantly toward achieving the goals of food security and environmental sustainability in India. In fact, by bridging existing gaps and capitalizing on emerging opportunities, smart agriculture can play a transformative role in the sustainable growth of India's agricultural sector.

Suggestions

To accelerate the adoption and effectiveness of smart agriculture technologies, several strategic recommendations are proposed. First, increasing awareness and providing targeted training programs are essential to enhance farmers' knowledge and skills in using these technologies effectively. This includes organizing workshops, demonstrations, and capacity-building initiatives tailored to the needs of diverse farming communities. Second, strengthening policy support by refining and enhancing existing frameworks is critical to ensure consistent and effective implementation of sustainable agricultural practices across regions. Policies must address regional disparities and focus on enabling equitable access to resources and technologies.

Promoting public–private partnerships can play a transformative role by fostering innovation, driving investment, and bridging gaps in technology dissemination. Collaborations between governments, the private sector, and community organizations can accelerate the development and scaling of smart agriculture solutions.

Sustained investment in research and development is vital to advance smart agriculture technologies and make them more accessible and affordable for farmers. Finally, addressing financial barriers is paramount, particularly for smallholder farmers, who often face high initial costs for adopting new technologies. Developing financial support mechanisms, such as subsidies, low-interest loans, and insurance schemes, can empower these farmers to embrace innovations and contribute to sustainable agricultural growth. These combined efforts are integral to creating a resilient, inclusive, and technology-driven agricultural ecosystem.

Conclusion

Smart agriculture and sustainable food systems present transformative solutions to the multifaceted challenges confronting Indian agriculture. The integration of advanced technologies such as precision farming, GPS-enabled machinery, and sensor-based irrigation systems represents a significant advancement in enhancing agricultural productivity and resource efficiency. These innovations not only boost yields but also contribute to more sustainable resource management and environmental protection. By leveraging these technologies, India can address critical issues such as water scarcity, soil degradation, and the impacts of climate change, thereby fostering a more resilient agricultural sector.

To fully realize the potential of smart agriculture, it is imperative to continue supporting these initiatives through effective policy frameworks, increased investment in research and development, and collaborative efforts among government bodies, private-sector stakeholders, and agricultural communities. Policies should be tailored to address regional disparities and facilitate broader technology adoption, while investment in research and development will drive further innovation and accessibility. Public–private partnerships can also play a crucial role in scaling these technologies and bridging gaps in implementation.

Furthermore, addressing barriers such as high initial costs and limited farmer awareness through targeted training and financial support mechanisms will be essential for achieving widespread adoption. Comprehensive strategies that integrate technological advancements with sustainable practices will not only enhance agricultural productivity but also contribute to long-term environmental sustainability and food security.

In conclusion, the study underscores the urgent need for a cohesive approach to advancing smart agriculture in India. By tackling existing challenges and capitalizing on emerging opportunities, India can pave the way for a more sustainable and productive agricultural future. This holistic approach will ensure that the benefits of smart agriculture are realized across all regions, leading to improved livelihoods for farmers and a stronger, more resilient agricultural sector.

References

Ahmad, F. (2023, September 19). Smart farming: Paving the way to food security and sustainability in India. *The Times of India*. https://timesofindia.indiatimes.com/blogs/voices/smart-farming-paving-the-way-to-food-security-and-sustainability-in-india/

Arpula, S., Gouthami, B., Nayak, B. B., & Shirisha, K. (2024). Smart irrigation systems in agriculture. In S. Sharma, J. Chauhan, Y. Mandela, A. T. Hemavathy, & H. S. Ballari (Eds.), *Insights into agriculture sciences* (Vol. II, pp. 11–26). Kripa-Drishti Publications. https://epubs.icar.org.in/index. php/IndFarm/article/view/54188

Arunachalam, A., & Ayyappan, S. (2013). Environmentally sustainable food systems for food security and nutrition. *Indian Farming*, *63*(7). https://epubs.icar.org.in/index.php/IndFarm/article/view/54188

Balkrishna, A., Pathak, R., Kumar, S., Arya, V., & Singh, S. K. (2023). A comprehensive analysis of the advances in Indian digital agricultural architecture. *Smart Agricultural Technology*, *5*, 100318. https://doi.org/10.1016/j.atech.2023.100318

Barooah, P., Alvi, M., Ringler, C., & Pathak, V. (2023). Gender, agriculture policies, and climate-smart agriculture in India. *Agricultural Systems*, *212*, 103751. https://doi.org/10.1016/j.agsy.2023.103751

Beulah, C., Sujni, P., Rajasree, R., & Sanjith, S. (2023). Effectiveness of organic smart agriculture and environmental sustainability in a post-pandemic world. *Indian Journal of Science and Technology*, *16*(37), 3090–3099. https://doi.org/10.17485/IJST/v16i37.1352

Dhanaraju, M., Chenniappan, P., Ramalingam, K., Pazhanivelan, S., & Kaliaperumal, R. (2022). Smart farming: Internet of things (IoT)-based sustainable agriculture. *Agriculture*, *12*(10), 1745. https://doi.org/10.3390/agriculture12101745

Kapoor, A., & Menon, H. (2023, December 2). Sustainable agriculture for the future: Insights from India's G20. *BW Businessworld*. https://www.competitiveness.in/sustainable-agriculture-for-the-future-insights-from-indias-g20/

Kapoor, S., & Pal, B. D. (2024). Impact of adoption of climate smart agriculture practices on farmer's income in semi-arid regions of Karnataka. *Agricultural Systems*, *221*, 104135. https://doi.org/10.1016/j.agsy.2024.104135

Kumar, V., Sharma, K. V., Kedam, N., Patel, A., Kate, T. R., & Rathnayake, U. (2024). A comprehensive review on smart and sustainable agriculture using IoT technologies. *Smart Agricultural Technology*, *8*, 100487. https://doi.org/10.1016/j.atech.2024.100487

Ranjan, S., Yadav, A., & Higginbottom, S. (2024). Precision agriculture: Harnessing technology for efficient farming. In S. Sengar & A. K. Singh (Eds.), *Fields of tomorrow: Advancements in modern agriculture* (p. 50). Kalyani Publishers. https://www.researchgate.net/publication/380531552_Precision_Agriculture_Harnessing_Technology_for_Efficient_Farming

Saikath, P. K., & Supriya. (2023). Advancing sustainable agriculture: A comprehensive review for optimizing food production and environmental conservation. *International Journal of Plant & Soil Science*, *35*(16), 3169–3181. https://doi.org/10.9734/ijpss/2023/v35i163169

Tanti, P. C., Jena, P. R., Timilsina, R. R., & Rahut, D. B. (2024). Enhancing crop yields and farm income through climate-smart agricultural practices in Eastern India. *Environmental Management*, *29*(35). https://doi.org/10.1007/s11027-024-10138-0

Chapter 5

Cashew Cultivation as a Driver of Sustainable Livelihoods in Southern India

Sarath Chandran M. C.[a], *Padma Rao Sahib*[b], *Krishnashree Achuthan*[c] *and Renju Chandran*[d]

[a]*Amrita School of Arts, Humanities and Commerce, Amrita Vishwa Vidyapeetham, Kollam, India*
[b]*Faculty of Economics and Business, University of Groningen, The Netherland*
[c]*Center for Cybersecurity Systems and Networks, Amrita Vishwa Vidyapeetham, Kollam, India*
[d]*School of Contemporary Knowledge Systems, Chinmaya Vishwa Vidyapeeth (Deemed to be University), Ernakulam, India*

Abstract

Cashew cultivation is vital for the promotion of sustainable livelihoods of smallholder farmers through a stable source of income while conserving the environment through agroforestry. The diverse impacts of cashew cultivation on smallholder farmers in southern India are assessed in this study regarding income enhancement, poverty alleviation, and the development of sustainable livelihood. This study uses the mixed-methods approach, where interviews as well as focused group discussions are conducted along with structured surveys on a randomized sample of 270 farmers. The findings highlight the fact that cashew farming greatly increases farm income for farmers, and yet it is hard to improve food access, educational access, social standing, and living standards. However, the study finds substantial concerns on long-term food security because monoculture cashew farming restricts crops production that are important for households, and volatile market prices for basic goods and services heighten economic vulnerability in farming communities. Such needs call for a friendly government policy that will take care of these problems. Key recommendations point to offering training programs, access to quality seed, and conducting best agricultural

Sustainable and Smart Farming Eco-systems:
Changing Landscapes in the Age of Technology, 59–73

Published under exclusive licence by Emerald Publishing Limited
doi:10.1108/978-1-83608-092-320261005

practices. Developing climate-resilient cashew varieties is also highlighted as essential for addressing environmental challenges and ensuring the long-term viability of cashew farming. Hence, this study fills the gap by offering valuable insights for policymakers and industry stakeholders and underlines the importance of an economic benefit derived from cashew cultivation that balances with potential drawbacks to sustainable livelihood.

Keywords: Sustainable livelihood; livelihood outcomes; cashew cultivation; India; smallholder farmers

1. Introduction

Cashew has become an important crop for sustainable agricultural growth, especially for smallholder farmers in India. It is a very resilient and high-value crop that makes a phenomenal impact on the economy in rural areas, especially because of regular income flow, building resilience, and bridging poverty levels at the grassroot level (Lukurugu et al., 2022). In the farming communities who have cash crops as their principal source of income, cashew tree cultivation makes an economy sustainable while offering environmental benefits under sustainable practices of agroforestry. Such integrated systems prove that cashew cultivation alone can balance economic viability with ecological stewardship.

Cashew farming has become transformative in southern India where agro-climatic conditions range (Habib et al., 2023). It allows smallholder farmers to utilize and optimize marginal lands while reducing dependency on traditional crops which are sensitive to variability in climate. This is pertinent in an age where environmental uncertainties are growing and the demand for sustainable agricultural practices is on the increase. Cashew trees are tolerant of drought and relatively adaptable; hence, they address many problems the farmers are faced with, ranging from erratic weather shifts to soil degradation (Melketo et al., 2019).

Socio-economic impacts cashew cultivation makes profound socio-economic contributions. Beyond providing a stable source of income, it increases the supply of education and healthcare and other essential services for farm households, thereby increasing their standard of living. In addition, the employment generated in the process of cashew processing and value addition makes ripple effects in supporting the broader rural economies (Hidayati et al., 2023). Women, who are widely engaged in the process of cashew processing, benefit extensively in promoting gender equity and financial independence.

On the environmental front, cashew production promotes sustainability through agroforestry systems that promote biodiversity and soil health (Olubode et al., 2018). Intercropping, in addition to ensuring diversified income streams for farmers, maintains ecological balance while cashew trees are companion planted with legumes, spices, or vegetables. These two practices illustrate how well cashew farming can fit into the global sustainability agenda: economically, socially, and environmentally (Pandey, 2024).

But in addition to this, the accelerated rate of expansion in cashew farming faces a challenge. The practice of monoculture contributes to a great risk for long-term food security and soil health, mainly because of the enhanced demand for cashew nuts in the market (Ayoo, 2022). Market price volatility exposes farmers to adverse financial risks without strong support systems. Interventions aimed at giving targeted solutions, such as intercropping activities, stable markets, and capacity-building initiatives among farmers, would solve these problems (Natarajan et al., 2022).

This chapter discusses the empirical data derived from the southern region of India and explores the multilevel effects of cashew farming in smallholder farmer systems. This chapter illustrates how cashew cultivation both generates socio-economic benefits, contributes to environmental considerations, and is accompanied by challenges. The study holds out the contention that its findings will provide actionable insights for the policymakers, stakeholders, and agricultural practitioners who seek to fully exploit cashew cultivation as a tool for sustainable rural development.

2. Cashew Cultivation and Sustainable Livelihoods

Cashew cultivation is rapidly emerging as one of the key 'backbone' drivers of socio-economic transformation and environmental conservation on the farm among smallholder farmers in rural India. The characteristic features of cashew farming – high value to markets, suitability for agroforestry, and adaptation to marginal lands – give cashew a crop pivotal significance in the realm of sustainable agriculture (Melketo et al., 2019). This section elaborates on the multifaceted contributions of cashew cultivations to livelihoods underlining both socio-economic and environmental dimensions.

2.1. Socio-economic Contributions

2.1.1. Income Diversification

One of the most attractive diversification opportunities in cashew is turning it into a steady source of income for smallholder farmers. Its perennial nature means that it can provide income throughout the year, in contrast to seasonal crops whereby there are growth cycles to be followed. Once established, cashews produce nuts annually, thus providing regular cash income to farmers. This economic stability reduces the risks that otherwise characterize traditional farming, such that farmers can withstand market fluctuation in other agricultural products.

In addition, the cashew value chain creates other revenue sources as value-added products. Even in processing cashews – cracking, roasting, and packaging – or converting cashews to butter, oil, and even cashew apple juice, business returns have the chance to be enhanced, as most value-added products sell at premium price both locally and overseas (Natarajan et al., 2022).

For instance, there is a case study from southern India, where cashew farming producers are said to have had an average increase of 40% in their annual income against producers who were still on cereals alone (Mc et al., 2024). This additional income enabled them to spend on upgrading their homes, better healthcare facilities, and quality education for their children.

2.1.2. Empowerment of Marginalized Communities

Cashew cultivation has been transformative for empowering marginalized groups, especially women and small-scale farmers. In cashew processing, women are heavily engaged in time-consuming tiring work that entails manual dexterity (Balarubini et al., 2017). The earnings they receive independently strengthen their status in the household and communities. Financial independence makes women recipients for discussion on family decisions such as education and financial planning. Programs supporting women participation in cashew processing have yielded great mileage. Examples of women's self-help groups in Kerala and Karnataka have together formed cooperatives to process and market cashew products. Apart from offering jobs, these enable them with leadership and entrepreneurial skills, thus reducing the gender gap in rural economies.

2.1.3. Community Development

The benefits of cashew cultivation stretch even beyond the household level to benefit the whole community. When farmers earn more, families will enjoy greater access to basic services such as healthcare and education. Cashew farming also assists in the development of infrastructure in rural areas in terms of processing units, transport networks, and storage facilities (Bansah et al., 2023). All this improves the quality of life in those rural areas.

Community-based organizations and cooperatives multiply the benefits. Through resource pooling, farmers have to access better markets and command higher prices for their produce. Improved farming techniques are also shared among them. These group efforts enhance social cohesion because of unity and shared progress among farmers.

2.2. Environmental Benefits

Cashew farming also contributes significantly to environmental sustainability. In agroforestry systems, the farming of cashew could take the place of major crops to achieve ecological stability and conservation of biodiversity, improvement of soil fertility, and other aspects for both agricultural productivity and broader environmental goals.

2.2.1. Soil Fertility

The cashew tree improves soil health due to its spread root systems, which curb soil erosion, especially in sloping terrains. The organic matter from fallen leaves also contributes to enriching the soil and thus lowers the need for synthetic fertilizers (Gajbhiye et al., 2020). Intercropping with legumes or other nitrogen-fixing plants means the cashew farm enhances the nutrient cycles, which keeps the agricultural ecosystem sustainable.

In Tamil Nadu studies, crops such as cowpea and green gram significantly enhanced soil nitrogen levels, improving yields both for the cashew trees and the intercrop. This approach represents yet another way cashew farming can be part of sustainable approaches to land management.

2.2.2. Biodiversity Conservation

Cashew farming, especially when done as an agroforestry system, supports biodiversity. Mixed-cropping with vegetables, spices, or other tree species establishes a heterogeneous habitat that will attract more birds, pollinators, and beneficial insects than a typical mono-species crop. This biodiversity does not only enhance ecological balance but also increases the other ecosystem services to farm users, such as natural pest control and pollination. The integration of cashew trees in mixed farming systems also minimizes the risks associated with monoculture. This is because a diverse farm landscape is strong against pests, diseases, and climatic variations and guarantees the long-term sustainability of such farmers.

2.2.3. Climatic Resilience

Cashew trees are best planted on marginal lands with severe climatic conditions. However, they are excellent crop products for regions experiencing drought and irregular rainfall. They thrive in poor soil conditions without requiring intensive irrigation, thus ensuring that farmers may also continue to grow them under undesirable environmental conditions.

As the climate change situation becomes unbearable, resilient crops like cashew will be of supreme importance. For example, in water-scarce districts of Andhra Pradesh, cashew has managed to be a hard nut for farmers to crack; it enabled them to sustain productivity even as water was declining. Cashew cultivation thus offers climate smart agriculture as an inherently imperative strategic approach.

2.3. The Double Advantage: Economic Benefits and Ecological Reliability

One aspect that sets cashew cultivation apart is how it can simultaneously target economic and environmental objectives. By offering stability in income to the poor, making them capable, and putting men and women in contributing positions, cashew farming contributes to better rural livelihoods. In turn, soil health, biodiversity, and climate resilience ensure that these objectives are long-term sustainable benefits. However, both benefits should be carefully managed with appropriate policies. Overemphasis on monoculture in response to the consistently high market demand for cashew nuts could undermine the environmental benefits of the agroforestry systems. Similarly, market price volatility exposes farmers to economic risks; hence, there is a need for policy measures to stabilize the prices and ensure fair returns from their produce (Natarajan et al., 2022).

3. Research Methodology

The present study applies a mixed-methods approach (Mc et al., 2024) to examine socio-economic and environmental implications of cashew cultivation in southern India. Use of data on both the quantitative and qualitative aspects ensures that research provides a comprehensive understanding of how cashew farming influences lives of smallholder farmers – to both measurable outcomes and experiences. This would provide sufficient evidence to inform policymakers and other stakeholders to design interventions that enhance the sustainability and profitability of cashew cultivation.

3.1. Quantitative Method

3.1.1. Survey Design and Implementation

A structured survey was developed to collect key quantitative indicators regarding the impacts of cashew farming. The tool used was the survey questionnaire that covered issues like income levels, availability of fundamental services, use of sustainable management practices, and the problems the farmers experience. Smallholder farmers totaling 270 were surveyed, with an equal representation of 90 from each state to attain balanced geographical distribution. Stratified random sampling was done so that farmers of different years of experience in cashew cultivation were included.

3.1.2. Data Collection

Income inequality: before and after entering cashew farming.
Healthcare access: food insecurity; education, etc.
Gains made through intercrop planting and organic farming adoption.
Challenges: price fluctuations; resources' accessibility, among others.

3.2. Qualitative Method

Qualitative methods were used to supplement the findings, further illuminating the lived experiences of farmers and other stakeholders. In-depth interviews were conducted with 20 farmers experienced in cashew farming and others who are recent adopters of cashew farming. Five were agricultural experts who offered a technical viewpoint on farming practices and market dynamics. And the remaining five were policymakers from the government who discuss government initiatives and problems of policy implementation.

Qualitative data were analyzed thematically to identify recurring patterns and unique insights. This approach enabled the study to highlight personal narratives and collective community perspectives that enriched the quantitative findings.

4. Insights from Qualitative and Quantitative Data

Qualitative and quantitative findings were combined to give a panoramic overview of the economic, social, as well as environmental impacts of cashew farming, such as the challenges of smallholder farmers. They show the multifaceted effects of cashew farming; areas of possible intervention are highlighted.

4.1. Economic Impacts

Cashew farming improved the economic livelihoods of smallholder farmers, from the quantitative data and the qualitative narratives. The adoption of cashew farming has allowed farmers to raise annual incomes by 40% as the average earning went up from ₹80,000 to ₹1,20,000. Income increase facilitated easier access to services, so that 68% of the farmers said that their children can be sent to school, and 54% said they enjoyed better healthcare. Additional earnings also increased living standards and therefore resilience to shocks in the long term.

Personal testaments highlighted the potential of cashew farming in changing individual's lives. A woman from Karnataka said:

> Through the income generated from processing cashews, my respect within my family has been enhanced.

These testimonials capture how economic empowerment extends beyond financial gain because it brings self-esteem and social recognition, especially among vulnerable groups of people.

4.2. Social Empowerment

The social impacts of cashew cultivation have been highly significant, mainly in terms of contributing to the empowerment of women and improving community integration. It was observed that 48% of cashew processing women achieved greater financial autonomy, which enabled them to take a larger part in decisions within the household. In addition to this, 67% of farmers were members of cooperatives: membership helped the members gain access to markets, enhanced their bargaining capacity, and shared resources among themselves.

Community-level interactions strengthened the cooperative model, thus encouraging social cohesion. In an Focused Group Discussion (FGD) conducted in Kerala, a respondent commented:

> The cooperative model has enabled us to share resources and negotiate better prices for our produce.

This collective approach not only improved market access but also encouraged solidarity among farmers to deal with common problems.

4.3. Environmental Contributions

Cashew farming has, through activities such as intercropping and enhancing biodiversity, contributed significantly to environmental sustainability. Farmers engaging in intercropping report real tangible environmental changes, as evidenced by the 72% increase in soil fertility after three years and a 58% increase in pollinators and beneficial insects on their cashew farmlands. These results have demonstrated the ecological benefits of growing various crops combined with cashews.

The farmers' experiences further brought environmental changes out vividly. A Tamil Nadu farmer said:

> The biodiversity on our farm has improved since we started inter-cropping cashews with vegetables and legumes.

Such practices not only enhance soil health but also create a balanced ecosystem, which contributes to long-term agricultural sustainability.

4.4. Challenges

Despite its benefits, cashew farming faces big challenges that hinder its potential. The dependence on monoculture laid forth issues such as 62% farmers fully depending on cashew for survival with implications on food security and soil degradation. Another issue of market volatility accounted for 59% where the lack of stable prices was termed a significant challenge and needed an early resolution process.

The voice of farmers was elicited through personal stories: Cuddalore's (a district of Tamil Nadu) farmer said:

> We need government intervention to stabilize prices and support during poor harvests.

These narratives underline the need for strong policies and infrastructures to handle market risks and enhance the resilience of farmers.

4.5. Analytical Framework: Sustainable Livelihood Framework (SLF)

Applying the SLF (Scoones, 1998, 2015) would analyze findings systematically. This framework assesses the interplay between five kinds of capitals, viz., human, natural, social, physical, and financial, while evaluating the sustainability of livelihoods derived from cashew farming (Mc et al., 2024; Natarajan et al., 2022).

5. Findings of the Study

Cashew farming has promising and challenging opportunities toward the benefit of farmers, communities, and the environment. This section covers positive impacts and challenges that transform cashew farming, henceforth pinpointing the aspects needing intervention for sustainable development (refer Table 5.1).

Table 5.1. Sustainable Livelihood Framework.

S. No.	Type of Capital	Findings of the Study
1	Human capital	Cashew farming has empowered communities, yet 48% of farmers acknowledged needing technical training in agroforestry and sustainable practices. Those dimensions must be addressed to achieve maximum productivity and ecological benefits.
2	Natural capital	The intercropping practices increased soil health remarkably, with farmers estimating 20% decrease in soil erosion in sloped farms. Such farming practices also promote biodiversity conservation, thus strengthening ecological sustainability of cashew farming.
3	Social capital	Cooperative approaches assisted 67% of the farmers with accessing resources and negotiating favorable market terms. On the other hand, 33% could not join due to steep membership fees, therefore, cooperative structures need to be more inclusive for all farmers.
4	Physical capital	Inadequate infrastructure remains a bottleneck, with 45% of farmers citing limited access to processing facilities and storage infrastructure. These shortcomings result in post-harvest losses and reduced profitability.
5	Financial capital	Farmers' savings increased by an average of 30%, reflecting improved financial security. However, persistent price volatility continued to undermine long-term financial stability, emphasizing the need for intervention in market mechanisms.

Source: Authors' compilation.

5.1. Positive Impacts of Cashew Cultivation

Cashew cultivation is now a multiple use driver for rural development, providing smallholder farmers with diverse economic, social, and environmental benefits. Its adaptability to semi-arid regions and potential integration with sustainable agriculture systems enhance its attractiveness as an alternative livelihood in areas where other forms of farming are not possible. Beyond its economic impacts, cashew farming has ripple effects in improving food security, empowering marginalized groups, and reaps ecological sustainability. The next section delves into the wider benefits of cashew farming to generate income, increase food security, and strengthen social empowerment in the pursuit of ecological gains.

5.1.1. Income Generation

Its contribution to household income is the greatest opportunity of cashew farming. The findings show that farmers' incomes average 40% in excess after adopting cashew cultivation, with the annual rise from ₹80,000 to ₹1,20,000. This higher income creates the possibility of saving for farming families and utilizing this money in various services such as education, healthcare, and housing – hence almost leading to improved standards of living and poverty alleviation. In addition, the steady income from cashew cultivation provides a stable source of income, especially where other agriculture-based activities are not feasible due to climatic or soil characteristics.

5.1.2. Enhanced Food Security

Though cashew is not a basic food crop, its economic benefits will indirectly ensure improved food security for farming households by enabling farmers to buy a variety of nutritional food products while rarer cases of hunger and malnutrition prevail. About 75% of farmers mentioned that supplement income from cashew farming improved their diet quality, and they can afford a high protein diet, such as lentils, eggs, and dairy products. This transition not only enriches the nutrition of the household but also better health outcomes in agricultural communities.

5.1.3. Social Empowerment

Cashew farming has been seen as an empowerment vehicle for women and others. Women take leadership roles in cashew processing and marketing, giving them economic independence and social status. In this study, 48% of the women involved in cashew farming activities reported gaining improved financial autonomy to contribute more actively to household decision-making processes. This empowerment leads women away from traditional roles to become active contributors to family income, for which they are rewarded. In addition, cooperative models have also helped to build social cohesion among farmers. They collectively bargain and share resources.

5.1.4. Ecological Gains

Cashew agriculture has been friendly to the environment, especially through techniques such as agroforestry which have helped integrate cashews and other agricultural crops in a symbiotic relationship in search of ecologic health without absolute farm productivity. Of the farmers practicing intercropping, 72% experienced improved soil fertility and 58% observed increased biodiversity, including pollinators and beneficial insects. Cashew trees, in addition to their natural resistance to drought, act as a climatic variability buffer, making them well suited for semi-arid regions. Another way that legumes, among other N-fixing crops, enhance soil health is by reducing synthetic fertilizer dependency as part of intercropping systems.

5.2. Cashew Cultivation – Challenges

While cultivating cashews is of great advantage to smallholder farmers, challenges abound that put the sustainability and long-term viability of cashew cultivation in risk. Primarily, overdependence on cashew monoculture has led to economic vulnerability due to volatility in the market, while its unfavorable practice has resulted in environmental degradation. Critical threats come from these areas that jeopardize not only livelihoods but also have broader implications for food security and ecological well-being. Therefore, the essential issues that cashew farmers face are risks in monoculture practice, market price, and environmental issues; however, the need for sustainability and policy interventions toward these issues highlights all the dire issues being faced and evaluated for decision-making.

5.2.1. The Risk of Monoculture

The dominance of cashew monoculture threatens crop diversity and long-term food security. Cashew farmers who wholly depend on monoculture are exposed to market crashes, pest attack, and climatic shocks. This monoculture minimizes the resilience of farming systems due to soil nutrient depletion and ecological balance disruption. This study determines that 62% of farmers relied solely on cashew farming. The study warned that these farmers would be highly susceptible to economic and environmental shocks. However, risk mitigation may occur through diversified intercropping and, therefore, livestock integration, which would make the farming model more sustainable and resilient.

5.2.2. Market Volatility

Cashew farmers are economically vulnerable mainly because of price volatility in the market and increasing costs of inputs; 59% of farmers reported difficulties in managing the unpredictability of cashew prices, which limits their capacity to plan long-term investments. These fluctuations often force farmers to rely on informal credit, which comes with high-interest rates and exacerbates financial stress. Without stable pricing mechanisms or government support, farmers struggle to achieve consistent profitability, making it challenging to scale their operations or invest in value-added processing.

5.2.3. Environmental Concerns

Although cashew trees are naturally resistant to drought, intensive growing of cashew trees has negative impacts on the environment unless the growing process is sustainable. The discussion raised alarm over soil nutrient depletion, especially if cashew is intercropped or cropped with another crop since, in the long term, fertility of the soil diminishes and the susceptibility to erosion increases risks to productive agriculture for decades ahead. In addition, the use of chemical fertilizers and pesticides in some farming systems contributes to pollution and biodiversity loss. These issues require a concerted action by conserving and adopting sustainable farming practices, including organic farming, intercropping, and integrated pest management.

6. Policy Implications

A multiple pronged policy approach would result in the solutions of the identified challenges and the maximum benefit from cashew farming. The following section deals with necessary policy inputs for growth in the cashew sector in an environment, which is supportive of sustainability and inclusivity.

Government support schemes must be instituted to directly tackle the resource gap challenges facing smallholder farmers. Subsidy on high-quality cashew seeds and planting materials ought to be presented to enhance yields and, therefore, profitability. Intercropping should also be promoted – this is one of the best practices for biodiversity improvement, soil health enhancement, and reduced risks in monoculture farming. Such subsidies may also be supplemented with educating farmers through awareness campaigns on advantages available to sustainable practices over the long term.

Another significant concern of the cashew growers is price volatility. For this purpose, the government should announce Minimum Support Price (MSP) for cashew nuts to ensure a fair return to the growers even during market shed periods. Direct market access through farmer cooperatives or digital platforms can reduce interference by intermediaries and add to the margins while making prices more transparent. Investment in storage and processing infrastructure would stabilize the supply chain and minimize post-harvest losses. The importance of training farmers needs to be placed since knowledge acquisition empowers farmers in the utilization of sustainable agriculture practices. Training modules could focus on:

> *Organic farming*: Decrease chemical fertilizers and pesticides used to maintain healthy soil and ecological integrity.
>
> *Agroforestry*: Combine cashew growing with other crops and trees to increase biodiversity and agricultural resilience.
>
> *Climate-resilient agriculture*: Training farmers on practices that help reduce climate risks, for instance, drought-tolerant crops and water harvesting systems.

These programs can be carried out through agricultural extension services, farmer field schools, and linkages with research institutions.

7. Recommendations

7.1. Promote Intercrop Systems

Intercropping is a good alternative for monoculture, as it solves the risk associated with one crop species, besides ensuring food security, soil health, and biodiversity. By growing intercrop cashew with cashew trees, the farmer has diversified sources of income and reliance on only one crop. Some examples of very good intercrops are legume vegetables and spices, which grow well under the cashew canopy and have a good effect on soil fertility through nitro-fixation. Technical support and incentives should be directed toward farmers for promoting intercropping systems by developing appropriate policy interventions.

7.2. Climate-resistant Varieties Development

Climate change impacts, such as prolonged droughts, erratic rainfall, and pest infestations, are severe threats to cashew farming. The research institutions should concentrate their efforts in developing a climate-resilient cashew variety that can be drought tolerant, pest resistant, and possibly extreme weather conditions tolerant. Accelerated development and dissemination of these varieties for farmers' use will come through collaborative efforts from research bodies, government agencies, and international organizations.

7.3. Strengthening Cooperative Models

The cooperative model could be shown to empower farmers through resource pooling, better market access, and improvement in bargaining power. Strengthening existing cooperatives and forming new ones, then, should be a priority for policies. Cooperative structures can also serve as easy platforms for problem-solving capabilities, dissemination of training skills, and sharing of resources among members. Financial support towards membership fees reduction and cost of operation would serve to make cooperatives more open so that small and marginal farmers benefit equally.

7.4. Integrating Social Welfare Programs

The linkage of agricultural policies with social welfare programs can work for and benefit the well-being of farming communities at large. For example:

Health and Nutrition: School-feeding programs can include cashew-based snacks, thus providing local market opportunities to farmers alongside nutritious childcare.

Education: Scholarships or subsidies for children of farmers operating under cashew cooperatives or government programs.

Community Infrastructure: Investments in healthcare centers, schools, and transportation in cashew-growing regions that would improve the quality of life for the farming family. These types of integrated programs would address the immediate needs of the community, while on the other hand would promote a symbiotic relationship between farming and rural development.

8. Conclusion

Cashew production proves to be a transformative livelihood for smallholder farmers with influential economic, social, and environmental benefits. This chapter elaborates on the way cashew farming enhances income generation, better food security, social empowerment, and ecological sustainability. The farmers have significantly increased the household incomes, which have facilitated investments in education, health, and nutrition. Socially, cashew farming has promoted gender balance since more women engage in processing and marketing. The cooperatives have made community solidarity much stronger. Environmentally, as shown above, cashew farming promotes biodiversity and enhances soil fertility through the intercropping system. All these show the potential of cashew farming as a future model for sustainable rural development.

These risks, such as monoculture risks, market price volatility, and environmental degradation, pose significant challenges to sustaining cashew farming over the longer term. Targeted interventions will be necessary to promote these ultimate goals of cashew cultivation. Policy support is required through minimum support prices, subsidies to promote sustainable or better practices of farming, and investment in processing infrastructures. Equipping farmers with training programs on agroforestry, organic farming, and climate-resilient techniques is key to increasing resilience and productivity (Pandey, 2024). With this multistakeholder approach, the economic benefits of cashew farming will be optimized, and the livelihoods of the farmers protected against predictable economic and environmental uncertainties.

Another potential area for future research could include addressing issues in cashew farming and its "spill over effects" on the rural economy and its possible linkage with value-added industries like cashew-based products and agro-tourism. Longitudinal studies may be done to analyze the long-term effects of intercropping and agroforestry practices about soil health and biodiversity. Comparative studies across regions or countries with similar climatic conditions may also be necessary for a fair comparison on replicating the success in other locations (Pandey, 2024). Such studies, then, would provide valuable evidence toward policymakers and other stakeholders so that cashew cultivation can continue to foster sustainable development on a broader scale.

References

Ayoo, C. (2022). *Poverty reduction strategies in developing countries*. IntechOpen eBooks. https://doi.org/10.5772/intechopen.101472

Balarubini, M., Karthikeyan, C., & Sujeetha, T. (2017). Empowerment of disaster affected farmers – A comprehensive study on developing feasible strategies for empowering thane affected cashew growers in Cuddalore district of Tamil Nadu, India. *International Journal of Current Microbiology and Applied Sciences*, *6*(11), 4001–4011. https://doi.org/10.20546/ijcmas.2017.611.469

Bansah, K. J., Arthur-Holmes, F., & Assan, E. (2023). Climate induced transformation of agriculture to artisanal mining economy in dry regions. *Journal of Rural Studies*, *99*, 11–19. https://doi.org/10.1016/j.jrurstud.2023.02.005

Gajbhiye, R. C., Pawar, S. N., Zote, V. K., & Sawant, B. N. (2020). Intercropping of different vegetable crops in new cashew plantation under Konkan conditions of Maharashtra. *International Journal of Current Microbiology and Applied Sciences*, *9*(6), 448–455. https://doi.org/10.20546/ijcmas.2020.906.059

Habib, N., Ariyawardana, A., & Aziz, A. A. (2023). The influence and impact of livelihood capitals on livelihood diversification strategies in developing countries: A systematic literature review. *Environmental Science and Pollution Research*, *30*(27), 69882–69898. https://doi.org/10.1007/s11356-023-27638-2

Hidayati, D. R., Garnevska, E., & Childerhouse, P. (2023). Enabling sustainable agrifood value chain transformation in developing countries. *Journal of Cleaner Production*, *395*, 136300. https://doi.org/10.1016/j.jclepro.2023.136300

Lukurugu, G. A., Mwalongo, S., Kuboja, N. M., Kidunda, B. R., Mzena, G., Feleke, S., Madeni, J. P., Masawe, P. A., & Kapinga, F. A. (2022). Determinants of adoption of enhanced cashew production technologies among smallholder farmers in Mtwara region, Tanzania. *Cogent Food & Agriculture*, *8*(1). https://doi.org/10.1080/23311932.2022.2137058

Mc, S. C., Chandran, R., Sahib, P. R., & Achuthan, K. (2024). Sustainable livelihoods through cashew cultivation: Insights from smallholder farmers in the southern region of India. *Discover Sustainability*, *5*(1). https://doi.org/10.1007/s43621-024-00558-y

Melketo, T. A., Geta, E., & Sieber, S. (2019). Understanding livelihood diversification patterns among smallholder farm households in Southern Ethiopia. *Sustainable Agriculture Research*, *9*(1), 26. https://doi.org/10.5539/sar.v9n1p26

Natarajan, N., Newsham, A., Rigg, J., & Suhardiman, D. (2022). A sustainable livelihoods framework for the 21st century. *World Development*, *155*, 105898. https://doi.org/10.1016/j.worlddev.2022.105898

Olubode, O., Joseph-Adekunle, T., Hammed, L., & Olaiya, A. (2018). Evaluation of production practices and yield enhancing techniques on productivity of cashew (Anacardium occidentale L.). *Fruits*, *73*(2), 75–100. https://doi.org/10.17660/th2018/73.2.1

Pandey, S. (2024). *Fiscal incentives for green growth: A mixed-methods study of the state of agroforestry and tree cover in India*. https://digitalcommons.dartmouth.edu/environmental_studies_senior_theses/5/

Scoones, I. (1998). *Sustainable rural livelihoods: A framework for analysis*. IDS eBooks. https://opendocs.ids.ac.uk/opendocs/bitstream/123456789/3390/1/Wp72.pdf

Scoones, I. (2015). *Sustainable livelihoods and rural development*. Practical Action Publishing eBooks. https://doi.org/10.3362/9781780448749

Chapter 6

Mainstreaming Climate Smart Agriculture (CSA) Strategies and Policies Within Financial Institutions' (FIs) Operations

Janardhana Anjanappa[a] *and Shridhar M. Samant*[b]

[a]*Climate Investment Advisory Group, TYCHE Investments, India*
[b]*Rishihood University, Rashtram School of Public Leadership (RSPL), Rishihood Foundation Program (RFP), Systems Thinking Lab (STL), India*

Abstract

This chapter proposes a framework for mainstreaming climate-smart agriculture (CSA) strategies within the operations of financial institutions (FIs), focusing on how FIs can support CSA to enhance agricultural productivity, resilience, and sustainability. The research explores the challenges posed by climate change to agriculture, with a particular focus on India. The study follows a multidimensional approach, starting with a review of global and regional CSA frameworks to identify best practices. It includes an institutional analysis. Stakeholder engagement to provide insights into barriers and opportunities for CSA adoption. The findings reveal that while Indian FIs have made efforts to invest in agriculture, the mainstreaming of CSA-related financing initiatives remains inconsistent. A critical gap exists in integrating CSA strategies into institutional policies and operations. The research emphasizes the need for financial mechanisms like green bonds and concessional loans to support CSA adoption. It also highlights the importance of establishing a robust monitoring, reporting, and verification (MRV) framework to track investments and ensure accountability. The study proposes a framework based on five principles: Commit, Manage, Promote, Improve, and Account. These principles guide FIs in embedding CSA strategies into their operations, risk management, and financial products. By aligning with CSA objectives, FIs can mitigate climate risks, unlock investment opportunities, and promote sustainable agricultural development. Mainstreaming CSA in FIs can significantly

Sustainable and Smart Farming Eco-systems:
Changing Landscapes in the Age of Technology, 75–88

doi:10.1108/978-1-83608-092-320261006

enhance the resilience of the agricultural sector, particularly in developing countries like India.

Keywords: Climate-smart agriculture; financial institutions; agriculture; climate change; risk management; sustainable development; India

1. Introduction

Climate change poses a significant threat to global economic systems, particularly in developing nations, where agriculture remains the backbone of livelihoods and food security. Agriculture is among the most vulnerable sectors to climate change, adversely impacting rural poverty, food systems, and socio-economic stability (Vermeulen et al., 2012). Furthermore, agriculture accounts for 19%–29% of global greenhouse gas (GHG) emissions, making it a critical sector for implementing climate mitigation strategies (Jarvis, 2014). As climate change intensifies, there is an urgent need for innovative approaches, such as CSA, to enhance resilience, improve productivity, and reduce emissions.

The Food and Agriculture Organization (FAO, 2009) has projected a requirement of \$9.2 trillion in investments in the agriculture sector from developing countries to limit global temperature increases below 2°C by 2050. This highlights the necessity for FIs to play a pivotal role in mobilizing and directing investments toward CSA. The decisions made by FIs have the potential to drive climate-resilient agricultural development, yet the progress in financing CSA-related interventions remains inadequate (UNFCCC, 2014). Globally, institutions like the Global Environment Fund and Climate Investment Funds (CIF, 2014) have initiated efforts to fund climate-friendly projects. However, these efforts are often limited to supply-side investments, with insufficient focus on demand-side efficiency projects.

In India, agriculture contributes less than 20% to the gross domestic product (GDP) but supports over half of the workforce (National Bank for Agriculture and Rural Development (NABARD), 2017). This sector is particularly susceptible to climate hazards, disproportionately affecting women and children due to their socio-economic vulnerabilities (United Nations Development Programme (UNDP), 2017). Although Indian banks and institutions have undertaken initiatives to scale up agricultural investments, the lack of cohesive strategies for mainstreaming CSA financing limits the impact of these efforts (Singh & Naik, 2017). Private-sector FIs also show minimal engagement in CSA, citing challenges like high upfront costs and limited understanding of CSA's long-term benefits (Schaafsma & Bell, 2018).

To address these gaps, there is a pressing need to integrate CSA strategies and policies into the core operations of FIs. This involves aligning institutional objectives with national and global climate goals while leveraging resources to scale CSA practices at the national level. Establishing robust CSA frameworks within FIs can enhance governance, foster innovation in financial products, and unlock funding opportunities from diverse sources (Corner-Dolloff et al., 2015). Moreover, mainstreaming CSA can mitigate climate risks and foster sectoral development, benefiting rural communities and ensuring food security.

This study proposes a framework for embedding CSA practices into FIs' operations. By streamlining CSA considerations into institutional strategies, it is possible to bridge existing policy gaps, create synergies across stakeholders, and build resilient agricultural systems. Furthermore, aligning financial operations with CSA goals will not only enhance economic growth but also contribute to achieving sustainable development objectives (Vermeulen et al., 2012). This approach underscores the critical role of FIs in addressing climate challenges and scaling CSA practices to ensure equitable and sustainable agricultural progress.

2. Scope of Mainstreaming CSA in the Indian Context

According to UNDP, climate change adaptation and poverty reduction can be addressed simultaneously by a strong enabling environment, empowered communities, private-sector engagement, and support to the development of institutions and policies. Hence, it is possible to achieve climate resilience in agriculture and reducing agricultural emissions by scaling up investments in Climate Smart Investments (CGAIR). Learning lessons from developing countries suggest that integrating climate change risks into poverty reduction at subnational level is an effective strategy for making poor farmers more resilient against climate change. Women and children are particularly vulnerable to climate change risks due to limited awareness and adoption of CSA practices, which further increases their exposure to climate-related shocks (Aryal et al., 2020).

The economic growth and FI lending have a positive correlation in the economy. An inclusive finance will provide better banking services at a reasonable cost by the variety of institutions for low-income and micro-, small-, and medium-sized enterprises. As far as interventions for farm credits are concerned, it comes under priority sector lending. The initiatives of establishment of Regional Rural Banks in 1976 and NABARD in 1982 have strengthened the cooperative structure and expansion of the cooperative network. However, there is a need for a multi-agency approach with a focus on agriculture. All in all, a new road map is required for the farmer credit mechanism (Belavadi et al., 2017).

In this context, the self-help groups' (SHGs) initiatives in India is the largest social mobilization initiative in the world with a market share of 8–9 million SHGs being supported by various nongovernmental organizations (NGOs), governments, and banks. It has become a channel to address poverty as an effort by the Reserve Bank of India, NABARD, and Government of India. Hence, the SHGs have been mainstreamed in the financial landscape, particularly Indian banking system and micro financial institutions (MFIs). The aggregated financial mobilization is much larger than multi-national corporations in the country. It has the potential to become an effective tool for solving social issues like poverty, and gender inequalities, etc. This has a great potential to increase the scope of various development initiatives on health, family planning, nutrition, sanitation, and agriculture. According to the National Rural Livelihoods Mission

(NRLM), it has the mandate to promote about 8–9 million SHGs across 640,000 villages over the next decade to support social mobilization and improve livelihood. Hence, SHGs have a huge potential to address issues related to agricultural productivity and improve the livelihood of poor households (HHs) in rural India (Ghosh, 2019).

Therefore, expanding the full range of practices and approaches among HHs, it is essential to streamline best practices like CSA within FIs' operations. At present, these practices are not systematically mainstreamed within these institutional operations. Therefore, a framework to mainstream these strategies and policies in FIs' operations is vital. As a result, streamlining CSA practices across FIs along with its financing will not only decrease the emissions from agriculture but also will increase the overall socio-economic impact from these investments.

3. Literature Review

CSA has emerged as a critical framework for addressing the intertwined challenges of climate change, food security, and sustainable agricultural development. CSA aims to sustainably increase productivity, enhance resilience, and reduce GHG emissions in agricultural systems, aligning with the goals of sustainable development (Corner-Dolloff et al., 2015). However, despite its potential, the implementation and scaling of CSA practices remain constrained by numerous socio-economic, biophysical, and institutional factors (Tanti et al., 2022).

Globally, the agricultural sector is responsible for approximately 19%–29% of GHG emissions, making it a significant contributor to climate change (Jarvis, 2014). Climate change exacerbates vulnerabilities in agriculture, particularly in developing nations, where it threatens rural livelihoods and food security (Vermeulen et al., 2012). South Asia, home to 40% of the world's poor, is particularly vulnerable, with estimates suggesting a 10%–50% reduction in agricultural productivity by 2050 if adequate measures are not implemented. Thus, CSA interventions are essential to mitigate climate risks while sustaining agricultural productivity.

Scaling CSA requires an enabling environment, including appropriate technologies, policies, and financial investments. Institutional synergies are critical for knowledge dissemination and capacity-building, especially among smallholder farmers, who are often the most vulnerable to climate impacts (FAO, 2013; Murugesan, 2024). FIs play a pivotal role in driving CSA adoption by providing necessary funding and financial instruments. However, current investments in CSA are insufficient, with many FIs lacking systematic frameworks to incorporate CSA into their operations (UNFCCC, 2014). Additionally, private-sector engagement in CSA financing is limited, often hindered by high upfront costs and uncertainties regarding returns (Schaafsma & Bell, 2018).

In India, CSA adoption faces unique challenges and opportunities. While agriculture contributes less than 20% to the GDP, it supports over half of the workforce, highlighting its socio-economic importance (NABARD, 2017). Indian banks, such as the NABARD, have initiated CSA-related projects but lack cohesive strategies to mainstream these initiatives across all institutions (Singh &

Naik, 2017). Furthermore, SHGs and MFIs have been instrumental in addressing rural poverty and gender disparities, providing a platform for scaling CSA interventions. However, their efforts are yet to achieve the desired impact due to fragmented approaches and limited policy integration (Government of India, Ministry of Rural Development, 2017).

Lessons from other regions underscore the importance of multi-stakeholder platforms and inclusive policymaking for CSA scaling. For instance, the Climate Smart Village approach in India demonstrates the potential of community-based CSA adoption but requires better alignment with institutional decision-making processes (Campbell et al., 2017). Additionally, global experiences highlight the need for robust MRV systems to evaluate the impact of CSA investments and foster accountability (Tyagi et al., 2014).

To address these gaps, integrating CSA into the operations of FIs is critical. By aligning institutional strategies with CSA objectives, it is possible to create a cohesive framework that supports knowledge sharing, financial innovation, and sustainable agricultural development. Such integration not only mitigates climate risks but also enhances socio-economic resilience, ensuring equitable growth in vulnerable regions (Neufeldt et al., 2015).

4. Approach and Methodology

To achieve the research objective of mainstreaming CSA strategies and policies within the operations of FIs, this study employs a structured, multidimensional approach (refer Fig. 6.1). The methodology begins with developing a conceptual framework by reviewing global and regional CSA frameworks, policies, and financing mechanisms to identify best practices and existing gaps. Institutional analysis will be conducted to examine the business strategies, policy commitments, and operational structures of Indian FIs to assess their alignment with CSA principles. Drawing on these insights, voluntary principles such as commitment, management, promotion, improvement, and accountability will be adapted to guide the integration of CSA into FI operations.

Stakeholder mapping and engagement form a critical component of this research. Key actors, including banks, policymakers, farmers, and NGOs, will be identified and engaged through interviews and focus group discussions to understand barriers, opportunities, and resource needs for scaling CSA interventions. Additionally, case studies of successful CSA financing models from India and other regions will be documented to provide replicable insights.

The study also emphasizes evaluating risks and opportunities associated with integrating CSA into FI operations. This will involve analyzing the climate risks and GHG emissions linked to FIs' existing agricultural portfolios and employing scenario modeling to assess the long-term benefits of CSA investments under varying climate and economic conditions.

Another key element is the design of tailored financial products and tools essential for mainstreaming CSA. Innovative financial instruments such as green bonds, concessional loans, and credit guarantees will be proposed to support CSA initiatives. Additionally, the study will explore incentive mechanisms like

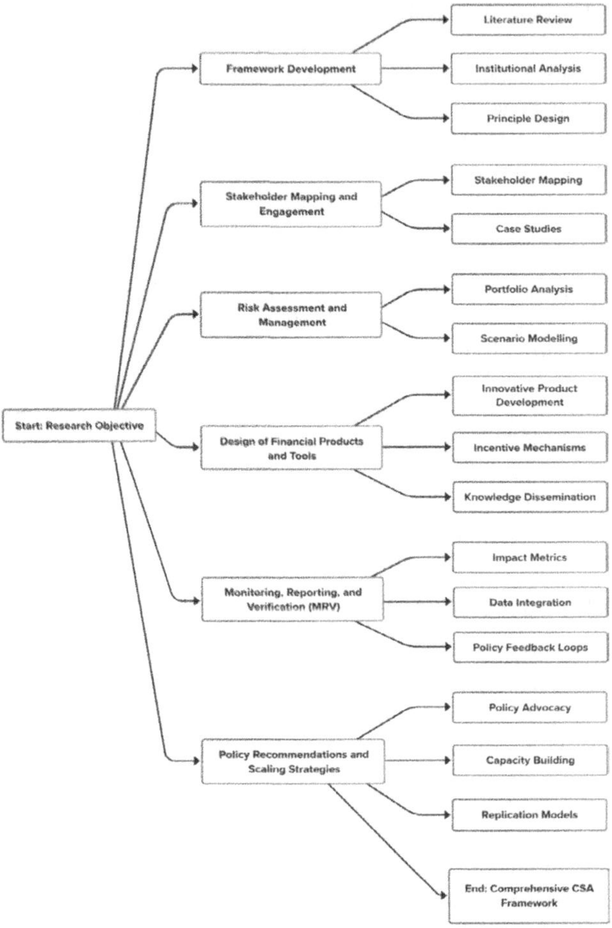

Fig. 6.1. Approach and Methodology. *Source*: Authors' compilation.

subsidies to encourage FIs and farmer adoption of CSA practices. Knowledge-sharing platforms will be developed to disseminate insights, tools, and best practices for accelerating CSA adoption.

To ensure accountability and measure impact, the research will develop a robust MRV framework. This includes establishing indicators to measure CSA integration, such as GHG reductions, resilience improvements, and socio-economic outcomes. Data integration through digital tools and geospatial technologies will enable comprehensive tracking of CSA financing and implementation. Policy feedback loops will be created to adapt CSA strategies continuously based on field data and policy outcomes.

The study concludes with actionable recommendations to scale CSA strategies and policies. It will advocate for engaging government and regulatory bodies to institutionalize CSA financing guidelines and propose capacity-building initiatives for FIs, farmers, and policymakers to strengthen understanding and implementation. Templates and models for replicating CSA strategies across FIs, regions, and agricultural sectors will be developed to facilitate widespread adoption.

By integrating these components, the research provides a comprehensive roadmap for embedding CSA into the operations of FIs. This approach not only aligns with global best practices but also addresses the unique socio-economic and climatic challenges faced by India, thereby unlocking opportunities for sustainable agricultural growth and enhanced climate resilience (Meinzen-Dick et al., 2013).

5. Proposed Framework and Development Plan

Table 6.1. Five Voluntary Principles for Mainstreaming CSA Within FIs' Operations.

Principle	Objective	Activity	Methodology	Outcomes
Commit	Assess institutional commitments	Assess senior management leadership, explicit strategies, policies, commitments, and targets	- Assess plans that integrate a systematic mainstreaming of CSA into business and strategic plans - Positions that are built in institutional structure promote and are accountable for meeting CSA objectives - Incentive structure, performance management systems, or other systems that track the delivery of systems at all levels	A CSA strategy to promote relevant strategic priorities, policy commitments, plans, and/or targets at an institutional level

(*Continued*)

Table 6.1. (*Continued*)

Principle	Objective	Activity	Methodology	Outcomes
Manage	Education on nonfinancial risk mitigation	Assess CSA risk of new investment, pipeline, or an existing portfolio	Assessing CSA risk of existing portfolio of projects. Developing approaches to assess and manage project risks	Identify the linkages with approaches for screening the new investments
Promote	Identify CSA considerations in routine business development	Share learning lessons and experiences to help further mainstream CSA considerations into activities and investments	Identify dedicated mechanisms that provide financing for projects, specific strategies to increase lending-related CSA	Promote approaches to generate instruments, tools, and knowledge. Encourage knowledge sharing and dissemination for creative and smart partnership between departments, thereby mainstreaming and implementation of investment at scale
Improve	Identify operational tools that improve the climate performance of activities	Identify operational tools and result framework	Assess the GHG impact of investments, CSA finance flows. Screen risk and opportunity for the same	Develop appropriate operational tools for tracking, monitoring and incorporate CSA considerations into day-to-day operations

Table 6.1. (*Continued*)

Principle	Objective	Activity	Methodology	Outcomes
Account	Transparency and reporting	Link CSA considerations in financial reporting	Identification of institutional reporting of CSA of investments and related policies and potential reporting on additional funds leveraged or mobilized	Improve overall institutional governance

Source: Bonnel & Swann (n.d.).

6. Discussions

Table 6.1 outlines a structured approach to embedding CSA strategies within the operations of FIs through five voluntary principles: Commit, Manage, Promote, Improve, and Account. These principles provide a comprehensive framework for aligning institutional goals with CSA objectives, addressing systemic challenges, and unlocking the potential of FIs to drive climate-resilient agricultural practices. Below is a discussion of each principle and its significance.

6.1. Commit

The principle of "Commit" emphasizes the need for FIs to institutionalize CSA strategies at the core of their operations. This involves assessing senior management leadership, strategic priorities, and explicit policies that promote CSA objectives. By integrating CSA into business and strategic plans, FIs can establish dedicated positions within their organizational structures to champion CSA goals. Incentive systems and performance management frameworks are also necessary to ensure accountability and sustained focus on meeting these objectives. The commitment phase is crucial for setting a strong foundation, as it ensures that CSA considerations are embedded into the organization's long-term vision and governance.

6.2. Manage

The "Manage" principle addresses the importance of understanding and mitigating nonfinancial risks related to CSA initiatives. FIs need to assess the risks associated with new investments, existing portfolios, and pipelines concerning climate vulnerabilities. This principle involves developing robust risk assessment tools and methodologies to evaluate CSA-related risks and opportunities systematically. Effective management enables institutions to identify areas where CSA can be integrated into their financial operations, reducing potential liabilities while enhancing the climate resilience of their portfolios.

6.3. Promote

Promotion focuses on fostering CSA adoption by embedding these considerations into routine business development activities. This includes creating dedicated financial mechanisms to fund CSA projects and strategies to increase lending tied to climate smart solutions. Sharing lessons learned and promoting knowledge dissemination among departments and stakeholders are vital for building momentum. By encouraging innovative financial instruments and partnerships, the "Promote" principle facilitates the scaling of CSA investments and integrates CSA into mainstream financial activities.

6.4. Improve

The "Improve" principle is centered on enhancing operational tools and frameworks to measure and optimize CSA activities. FIs need to assess the GHG impacts of investments and identify tools to track, monitor, and incorporate CSA considerations into day-to-day operations. Developing robust monitoring and evaluation systems ensures that CSA investments are achieving their intended outcomes, providing actionable insights to refine strategies. This principle ensures that CSA initiatives remain dynamic and responsive to evolving climate and financial conditions.

6.5. Account

Transparency and accountability are the focus of the "Account" principle. Linking CSA considerations to financial reporting allows FIs to track the intensity of CSA investments and policies systematically. By establishing clear reporting mechanisms, institutions can demonstrate their contributions to climate resilience and sustainable development. This principle not only enhances institutional governance but also builds trust among stakeholders, including investors, regulators, and beneficiaries.

6.6. Significance of the Framework

These five principles collectively provide a roadmap for FIs to systematically integrate CSA into their operations. They encourage institutions to move beyond fragmented and ad hoc approaches to a cohesive strategy that aligns with global climate goals. By adopting this framework, FIs can position themselves as key enablers of sustainable agricultural development while mitigating risks and seizing new opportunities.

The framework also emphasizes scalability and adaptability, allowing institutions to tailor their strategies to local contexts and priorities. Ultimately, the successful implementation of these principles can contribute to achieving national and global climate objectives, enhancing resilience across agricultural systems, and ensuring inclusive growth for vulnerable populations.

6.7. Comparison with Other Frameworks

6.7.1. UNFCCC's Financial Mechanism Framework

The UNFCCC's financial mechanism emphasizes investments in climate resilience and low-carbon development. Similar to the CSA framework, it includes principles for transparency, risk assessment, and stakeholder engagement. However, the CSA framework in Table 6.1 is more tailored to the agricultural sector and addresses specific challenges, such as the integration of smallholder farmers and the unique risks of agricultural portfolios. In contrast, the UNFCCC's framework often focuses on broader climate investments, such as renewable energy and infrastructure.

6.7.2. Climate Investment Funds

The CIF framework emphasizes long-term GHG mitigation potential and scalability. While it aligns with the CSA principles of "Promote" and "Improve," CIF often focuses more on supply-side investments, such as large-scale renewable projects, rather than demand-side efficiency measures prevalent in CSA. The CSA framework, on the other hand, explicitly addresses the inclusion of smallholder farmers and community-based resilience practices, making it more socially inclusive.

6.7.3. FAO's CSA Approach

The FAO's CSA framework broadly aligns with the principles in Table 6.1, as both aim to enhance productivity, resilience, and emission reductions. However, the FAO framework emphasizes policy-level interventions and farmer capacity-building, whereas Table 6.1 principles are institutionally focused, targeting FIs specifically. The "Account" principle, for instance, adds a layer of financial transparency and governance that is less emphasized in the FAO framework.

6.7.4. World Bank's Climate Action Framework

The World Bank's framework integrates climate action into financial systems through capacity-building, knowledge sharing, and financing tools. Similar to Table 6.1's "Promote" and "Improve" principles, the World Bank framework supports innovative financing mechanisms and operational tools. However, Table 6.1 places greater emphasis on embedding CSA into institutional governance through the "Commit" and "Account" principles, offering a more comprehensive approach for FIs.

The five voluntary principles in Table 6.1 provide a robust, tailored framework for mainstreaming CSA within FIs. Compared to broader frameworks like those of UNFCCC, CIF, FAO, and the World Bank, the CSA-specific framework emphasizes institutional integration, risk management, and financial transparency. Its focus on agriculture-specific challenges and stakeholder inclusion makes it particularly valuable for addressing the vulnerabilities of smallholder farmers and enhancing the socio-economic impacts of CSA investments. By aligning these principles with complementary elements from other frameworks, FIs can adopt a holistic approach to scaling CSA, fostering resilience, and achieving sustainable agricultural development (Saikanth et al., 2023).

7. Conclusions

This study highlights the critical importance of mainstreaming CSA strategies and policies within the operations of FIs. CSA presents a transformative approach to addressing the dual challenges of agricultural sustainability and climate change mitigation. By integrating CSA into FIs' operations, institutions can significantly enhance productivity, build resilience, and reduce GHG emissions in the agricultural sector, particularly in climate-vulnerable regions like India.

The research emphasizes that while Indian FIs have made notable efforts in promoting agricultural investments, there remains a significant gap in systematically embedding CSA strategies. The proposed five voluntary principles – *Commit*, *Manage*, *Promote*, *Improve*, and *Account* – provide a comprehensive framework to address this gap. These principles advocate for institutional commitment, robust risk management, innovative financial products, continuous operational improvements, and transparent reporting mechanisms. Together, they create a cohesive strategy for scaling CSA investments and enhancing the socio-economic impact of financial interventions in agriculture.

Moreover, the study underscores the necessity for multi-stakeholder collaboration, policy alignment, and capacity-building initiatives to overcome systemic barriers. By streamlining CSA into the financial landscape, FIs can unlock new funding opportunities, foster innovation, and contribute to achieving national and global sustainable development goals (Haldar & Pearlin, 2023).

7.1. Future Research Directions

While this study provides a foundational framework for mainstreaming CSA, several areas warrant further exploration:

1. *Assessment of CSA Impact on Financial Performance*: Future studies should investigate the long-term financial benefits for FIs that integrate CSA practices into their portfolios. Quantitative analyses of return on investments in CSA initiatives can strengthen the business case for CSA financing.
2. *Inclusion of Technological Innovations*: Research should explore how emerging technologies, such as artificial intelligence, blockchain, and geospatial analysis, can enhance the efficiency and scalability of CSA frameworks within FIs.

3. *Farmer-centric Approaches*: While institutional strategies are vital, future research should focus on the socio-economic outcomes for farmers adopting CSA practices. Understanding farmer behavior, adoption barriers, and localized impacts can further refine CSA frameworks.
4. *Comparative Analysis Across Regions*: Conducting comparative studies of CSA mainstreaming efforts in different regions, particularly within developing economies, can provide valuable insights into best practices and transferable strategies.
5. *Policy Synergies and Government Role*: Future work could explore the role of government policies in supporting FIs' CSA initiatives. Investigating the interplay between public incentives, regulatory frameworks, and private sector efforts can identify opportunities for synergy.

References

Aryal, J. P., Sapkota, T. B., Rahut, D. B., & Jat, M. L. (2020). Agricultural sustainability under emerging climatic variability: The role of climate-smart agriculture and relevant policies in India. *International Journal of Innovation and Sustainable Development*, *14*(2), 219–245.

Belavadi, V. V., Karaba, N. N., & Gangadharappa, N. R. (Eds.). (2017). *Agriculture under climate change: Threats, strategies and policies* (Vol. 1). Allied Publishers.

Bonnel, A., & Swann, S. (n.d.). *Mainstreaming climate action within financial institutions emerging practices*. The World Bank Group and Agence Francais de Development (AFD) on behalf of, and in collaboration with, the Africa Development Bank (AfDB), Asian Development Bank (ADB), CAF Latin American Development Bank (CAF), Credite Agricole S. A., Cassie des Depots Group (CDC), Development Bank of South Africa (DBSA), European Investment Bank (EIB), European Bank for Reconstruction and Development (EBRD), InterAmerican Development Bank (IADB), International Bank for Reconstruction and Development (IBRD), International Finance Corporation (IFC), Japan International Cooperation Agency (JICA), KfW Bankengruppe (KfW), and the Multilateral Investment Guarantee Agency (MIGA).

Campbell, B. M., Beare, D. J., Bennett, E. M., Hall-Spencer, J. M., Ingram, J. S. I., Jaramillo, F., & Shindell, D. (2017). Agriculture production as a major driver of the Earth system exceeding planetary boundaries. *Ecology and Society*, *22*(4), Article 8.

Climate Investment Funds (CIF). (2014). *Delivering at scale empowering transformation (annual report)*. Climate Investment Funds (CIF).

Corner-Dolloff, C., Nowak, A., & Lizarazo, M. (2015). Climate-smart agriculture investment prioritization framework.

Food and Agriculture Organization. (2009). *The investment imperative: Paper from the FAO high-level conference on world food security: The challenges of climate change and bioenergy*.

Food and Agriculture Organization. (2013). *Climate-smart agriculture sourcebook*.

Ghosh, M. (2019). Climate-smart agriculture, productivity and food security in India. *Journal of Development Policy and Practice*, *4*(2), 166–187.

Government of India, Ministry of Rural Development. (2017). *National Rural Livelihood Mission (NRLM): District-level functionaries development role manual*.

Haldar, N., & Pearlin, R. (2023). Sustainable and climate smart agriculture for food security: A review. *Journal of Experimental Agriculture International*, *45*(11), 229–239.

Jarvis, A. (2014). Rising to the challenge of establishing a climate smart agriculture.
Meinzen-Dick, R. S., Bernier, Q., & Haglund, E. (2013). *The six "ins" of climate-smart agriculture: Inclusive institutions for information, innovation, investment, and insurance*. CGIAR Research Program on Climate Change, Agriculture and Food Security (CCAFS) Working Paper.
Murugesan, R. (2024). Climate-smart agriculture in India: Greenhouse gas mitigation strategies. *Journal of Agrometeorology*, *26*(4), 526–534.
National Bank for Agriculture and Rural Development. (2017). *Annual report 2016–17*.
Neufeldt, H., Negra, C., Hancock, J., Foster, K., Nayak, D., & Singh, P. (2015). *Scaling up climate-smart agriculture: Lessons learned from South Asia and pathways for success* [ICRAF Working Paper No. 209]. World Agroforestry Centre. http://dx.doi.org/10.5716/WP15720.PDF
Saikanth, D. R. K., Kumar, S., Rani, M., Sharma, A., Srivastava, S., Vyas, D., Singh, G. A., & Kumar, S. (2023). A comprehensive review on climate change adaptation strategies and challenges in agriculture. *International Journal of Environment and Climate Change*, *13*, 10–19.
Schaafsma, M., & Bell, A. (2018). Scaling up climate smart agriculture: Lessons from ESPA research.
Singh, C., & Naik, G. (2017). *Financial inclusion in India: A case study of Gubbi*. SSRN.
Tanti, P. C., Jena, P. R., & Aryal, J. P. (2022). Role of institutional factors in climate-smart technology adoption in agriculture: Evidence from an Eastern Indian state. *Environmental Challenges*, *7*, 100498.
Tyagi, N. K., Joshi, P. K., Aggarwal, P. K., & Khatri-Chhetri, A. (2014). Institutions and policies to scale out climate-smart agriculture: South-South exchanges.
UNFCCC. (2014). Fifth review of the financial mechanism (technical paper): Standing committee on finance. UNFCCC.
United Nations Development Programme. (2017). *Human development report 2017*.
Vermeulen, S. J., Aggarwal, P. K., Ainslie, A., Angelone, C., Campbell, B. M., Challinor, A. J., Hansen, J. W., Ingram, J. S. I., Jarvis, A., Kristjanson, P., Lau, C., Nelson, G. C., Thornton, P. K., & Wollenberg, E. (2012). Options for support to agriculture and food security under climate change. *Environmental Science & Policy*, *15*(1), 136–144.

Chapter 7

Harnessing Technology for the Marketing of Agricultural Products in India

Varsha Khattri

FORE School of Management, New Delhi, India

Abstract

The revolutionary influence of technology in improving agricultural marketing in India is examined in this chapter. It starts with a summary of the current state of the agricultural marketing system, emphasizing the difficulties and inefficiencies that farmers encounter, including restricted market accessibility, opaque pricing, and poor infrastructure. The adoption of technology is then highlighted in relation with modern agriculture, emphasizing how it may be used to overcome these obstacles and raise profitability, sustainability, and production. This chapter looks at a number of digital tools that have been created to help farmers market their goods, such as mobile apps and e-commerce websites, which allow for improved supply chain management, direct market access, and real-time pricing information. Additionally, the use of digital marketing techniques, such as social media, search engine optimization (SEO), and targeted ads, is being looked into as a means of boosting agricultural product market visibility and customer involvement. In order to effectively utilize the advantages of technology in agricultural marketing, this chapter also delves on the major barriers to adoption of technology in agriculture. The way forward discusses the future, highlighting the necessity of infrastructure development, digital literacy, and legislative assistance. India's agriculture industry may become more sustainable, profitable, and efficient by incorporating technology, which will benefit farmers and customers together.

Keywords: Agricultural marketing; technology; digital platforms; digital marketing; agricultural products

Sustainable and Smart Farming Eco-systems:
Changing Landscapes in the Age of Technology, 89–102

doi:10.1108/978-1-83608-092-320261007

Introduction to Agricultural Marketing System in India

The Royal Commission on Agriculture (1928) recommendations, which were incorporated into the Model Bill of 1938, are the origin of agricultural market regulation in India. The Agricultural Produce Marketing Regulation (APMR) Act, which was created after India gained its independence, was a significant piece of legislation for governmental control over the country's agricultural sector. Many Indian states embraced the Agricultural Produce Marketing Committees (APMCs) in the 1960s when it gave state governments the option to create them (Saha et al., 2024). When APMCs were first established and for many years later, they were praised as a creative and democratic way to guarantee higher agricultural output prices through auctions and shield farmers from the high cost of marketing and produce loss (GoI, 2009). The physical market infrastructure, known as the APMC or *mandi*, is present in every Indian state with the exception of Jammu and Kashmir, Bihar, Kerala, and Manipur. They act as tangible organizations that control market processes like weighing, selling, grading, and payment methods.

The government is required to use traders and middlemen (commission agents) to purchase specific "notified" agricultural commodities from farmers at the APMCs. A commission is paid to the farmers by these APMC-licensed dealers and agents in return for making it easier for them to sell their produce. In addition, the farmers and dealers pay a market fee to the APMCs, which goes into building and maintaining the physical infrastructure. In order to prevent traders and mercantile capital from taking advantage of farmers, the APMCs were intended to serve as a platform for marketing activities. However, the APMC system witnessed a significant decline over time as a result of vested interests, narrow-minded thinking, and bureaucratic flaws. The market system's flexibility and competitiveness were jeopardized by the Essential Commodities Act and other regulations. A link between the political class and the APMCs' traders and intermediaries developed as a result of the state governments' practice of appointing their own candidates to the boards of the APMCs as they began to receive money from them. Additionally, the APMR required traders at the APMCs to hold a license, and in order for a license to be granted, the trader had to possess a store or go-down location inside the APMC's boundaries. This promoted rent-seeking behavior and created a significant obstacle to new business owners entering the market. Additionally, the authorized dealers and middlemen banded together to form groups. Farmers' ability to negotiate prices was weakened by these two variables, which also hindered the entry of new business owners (GoI, 2013).

As a result, throughout time, the system served the political–trader–middlemen nexus to advance their monopolistic tactics rather than promoting effective marketing strategies that benefited the farmers. The excessive marketing fees imposed on farmers and the inadequate maintenance of market infrastructure were the results of the APMCs' lack of competition.

As the demographics of the agricultural industry and its output change, there have been several stages in the marketing of farm products, ranging from bartering to internet marketing. The implementation of complete marketing strategies is getting increasingly difficult due to the diversity of regional, technological, attitude, economic, and political elements. As the population grows, so does the demand for high-quality food. Ensuring proper production and marketing on a worldwide scale is becoming more and more crucial. The Indian agricultural marketing system is facing new difficulties as a result of the rising agricultural output. There is an urgent need to strengthen agricultural marketing systems (Saraf et al., 2022). A number of issues confront the Indian agriculture sector, such as maintaining investment, climate change, price volatility, undeveloped markets, the profitability of agriculture in comparison to other industries, raising smallholder productivity, supporting a large population, and controlling expenses. Application of modern technology in the field of agriculture in India can resolve many of these issues efficiently and effectively.

Modern Technology in Agriculture in India

Over the years, India's agricultural landscape has seen substantial changes, and numerous new technologies have been created. The newest trends and solutions are also being used by traditional farmers and agro-tech startups to increase production along the food value chain. In order to boost farmer productivity and yield more crops, it involves implementing new technology like cloud-based solutions and other pertinent advanced agriculture management strategies.

In order to control infestations earlier and more precisely, grape producers in India have started to detect and geolocate crop diseases or plagues. Additionally, less damaging pesticides are used on the crop as a result. Many farmers utilize soil mapping tools to figure out how much fertilizer is best for their fields. Farmers are deploying drones, which enable more precise pesticide spraying.

Indian sugarcane farmers have begun utilizing technology to determine when it is best to harvest their crops, enabling them to better organize their harvest and increase yield. A few Indian farmers have also started predicting crop production, weather patterns, and *mandi* price trends using artificial intelligence (AI)/machine learning-powered technologies (Bhatt et al., 2024). To escape the whims of labor shortages, some farmers have also started experimenting with self-driving tractors and robots that plant seeds.

The AI for Agriculture Innovation (AI4AI) initiative was started by the World Economic Forum India Center for the Fourth Industrial Revolution in partnership with the state of Telangana and India's Union Ministry of Agriculture. Agri-inputs, consumer, food processing, banking, insurance, and technology companies, as well as startup ecosystems and farmer cooperatives, are among the organizations involved, reflecting the complexity of the situation. Workshops were conducted by the AI4AI program to learn how to give Indian farmers access to the AI technologies they need to increase their productivity and income. With both advising services, AI-based quality testing, and a digital marketplace to link buyers and sellers, the project revolutionized chili farming for thousands in India's Khammam area in Telangana. According to participating farmers, their revenue has doubled (Jurgens & Kaushik, 2024).

Increased production, increased investment, sector diversification, use of modern techniques, growth of horticulture and floriculture, increased export volume, and the development of the food processing industry are some of the emerging trends in the agricultural sector that are very noticeable in the post-liberalization era.

Recent Developments in Agricultural Technology

Agricultural Drone Technology

Drones are employed extensively in agriculture to enhance crop growth, maintenance, and cultivation techniques, as well as in the delivery of medical care and protection support. These aerial transporters, for instance, are used to assess crop conditions and carry out more effective fertilizing techniques to increase yields. Even the availability of hovering robots aids farmers in gathering data and conducting surveys over wide regions to improve their understanding of their fields. Drones have made it possible to remotely monitor crops and livestock more frequently and affordably. Additionally, it aids in the analysis of field conditions and the selection of suitable interventions, including pesticides, fertilizers, and nutrients.

Blockchain Technology

Given the rising demand for supply chain transparency, data management, food safety, and traceability in the global food ecosystem, blockchain, a game-changing technology that just upended global food chains will continue to reinforce the dynamics of agribusiness. From verifying the accuracy of data to certifying the origins of produce and recording transactions, blockchain has the power to alter agri trade while maintaining food safety and quality requirements (Xiong et al., 2020). Due to the use of smart agriculture tactics by several industry leaders, the technology is expected to grow quickly.

Aerial Imaging

Geographic information system technology is used in aerial photography to assess the potential of irrigation projects and their effects on drainage, erosion, and land degradation and extent of disease or infestation. The visuals generated by aerial imaging make it possible to evaluate the leaves of a single plant. In order to safeguard crops from environmental hazards, these images are actively employed to identify pests and diseases. It also assists farmers in keeping an eye on the state of their farms' soil and is helpful during the summer months when water supplies are at their lowest.

Internet of Things (IoT) in Agriculture

By installing a variety of sensors on farms, IoT is helping the agricultural industry in a transformative way. Light, humidity, soil moisture, temperature, crop health, etc. are all monitored by these sensors. The following are some of the main applications of IoT in agriculture:

- A variety of farm sensors, including wearables, button cameras, robots, control systems, and driverless cars, aid in gathering information for analyzing the farm's performance.
- Drones, both airborne and ground-based, are used for field analysis, irrigation, monitoring, and crop health evaluation.
- The application of instruments to forecast soil, humidity, temperature, rainfall, and other natural disasters.

Hydroponics and Vertical Farming

The idea behind hydroponic farming is to use less water while producing higher final product yields, textures, and tastes. Hydroponically grown plants don't require large root systems, which enables them to put more energy into producing leaves and fruits. These plants grow more faster and are more resistant to pests and other illnesses as a result of being grown indoors.

In terms of sustainability, vertical farming cuts down on transportation and harmful emissions by enabling farms to be situated close to or inside densely populated regions. Growing crops in urban settings is made possible by vertical farming, which also helps to make fresh foods more easily accessible. Compared to traditional farming methods, this farming drastically lowers the amount of land space needed to cultivate crops.

Digital Platforms for Farmers

Farmers can interact with value chain participants, customers, and markets using digital platforms and e-commerce solutions. These platforms give farmers access to market data, allow price discovery, and streamline direct marketing. Smallholder farmers can increase their business prospects and get past conventional market barriers by utilizing digital market access solutions.

In September 2024, Union Minister of Information and Broadcasting Ashwini Vaishnaw told reporters that the government intends to develop a mobile end-user transaction and information platform for farmers that is comparable to the Unified Payments Interface as part of the Digital Agriculture Mission, which will receive ₹2,187 crore in funding (Haq, 2024). The so-called agri stack, a farmers' database created by compiling data from the PM-KISAN cash-transfer program that has bank-authenticated, land-holdings-based records of almost 100 million farmers, would be used to build the digital platform. Every landholding farm household receives a cash handout of ₹6,000 a year under the scheme. This amount is paid in three equal installments of ₹2,000, one every four months. The portal will contain information that will allow agricultural businesses and creditors to provide faster services, including farmers' registries, land maps, weather updates, and crop yields for farm insurance.

Digital platforms provide a multitude of resources and flexibility. Geographical limitations are eliminated because farmers can access training materials, webinars, and seminars from any location. By facilitating ongoing education, these platforms help farmers stay abreast of the most recent developments and industry best practices. In the end, farmers' livelihoods and productivity are improved by this digital revolution (Ananda et al., 2023). The following are a few examples of digital platforms:

a. AgriApp – The mission of AgriApp is to enable and equip farmers with the next generation of farming techniques while bridging the gap between the agricultural ecosystem and the digital world. Turning every farmer into a Digitally Facilitated "Agri-Entrepreneur" to Deal with Issues with Food, Fiber, Fertilizer, Fuel, and Feed (AgriApp, 2024).
b. eNAM (National Agriculture Market) – In order to establish a single national market for agricultural products, the eNAM (2024), a pan-India electronic trade portal, links the current APMC mandis.
c. FarmERP – Since technology adoption began to drastically alter the agriculture sector, FarmERP has become the leader in farm business management. It is dedicated to increasing digital farming's total efficiency and establishing "sustainable agriculture" as the new standard. They use two of the sustainable development goals of the United Nations – Zero Hunger and Climate Change – as a guide to propel the agriculture sector's overall expansion. FarmERP (2024) has consistently worked to improve food sustainability and safety.
d. Jio Agri (Jio Krishi) – Introduced in February 2020, the Jio Agri (Jio Krishi) platform empowered farmers by digitizing the agricultural ecosystem along the whole value chain. The main purpose of the platform is to provide advice by using data from independent apps. Its advanced capabilities employ data from multiple sources, feed it into AI algorithms, and then provide accurate, personalized recommendations (Jain & Dhar, 2022).
e. DeHaat Farmer App – Over 1.4 million farmers across 12 major agri-based states in India can access a variety of services through the DeHaat Farmer App, including voice calls in regional languages, crop advisories, weather reports, local mandi prices, and periodic crop reminders. Agri-experts can provide farmers with immediate consultations via the app regarding crop health advice, chemical dosage, and the use of premium input goods. Convenient input and output product purchases and sales can also be made via the app (AgRevolution, 2024).
f. KRIBHCO-CSC-eGov – Common Services Centers (CSC) e-Governance Services India had collaborated with Krishak Bharati Cooperative (KRIBHCO) to provide access to fertilizers and agricultural inputs to farmers via 4 lakh village-level entrepreneurs (Chhapia, 2024).

Through the interventions and collaborations of numerous startups, nongovernmental organizations, government agencies, and village-level institutions such as farmer producer organizations and self-help groups, the impact of these digital apps is growing at a considerable rate. By engaging rural residents in production, consumption, or both, the businesses are expanding their market share and enhancing their prospects for a living (Ascent Group India, 2024). Thus, if businesses and entrepreneurs invest and seek out alternative solutions that can be embraced and depended upon in the long run, these new and potential markets have the ability to bring about a complete paradigm change.

Use of Digital Marketing in Agriculture Marketing

Identifying the farm products, producing them to the proper quality, storing them to guarantee their availability, and then moving them to the locations where they are needed are the steps involved in traditional agricultural product marketing. The primary means of communication between the producer and the consumer is nonverbal. The producer (farmer) creates a product that consumers want, makes it in a way that is thought to meet their needs, stores it, and distributes it to areas where there is thought to be sufficient demand.

The marketing of agricultural products has been completely transformed by digital marketing, which has given farmers and agribusinesses new and creative ways to connect with customers. This contemporary method uses a variety of internet channels and tactics to interact with customers, increase brand recognition, and boost sales. The following are some significant ways that digital marketing is changing the agriculture industry.

E-commerce and Online Marketplaces

The purchasing and selling of agricultural products and associated services via online channels is known as e-commerce. Fresh produce, processed foods, cattle, poultry, agricultural inputs (including seeds, fertilizer, and equipment), and agricultural services like consulting and advisory services are all included in this broad category of activities. The constantly changing digital landscape and the growing need for easy access to agricultural products have propelled the e-commerce of agricultural products market's recent notable expansion. Fresh fruit, cereals, animals, and agricultural equipment are just a few of the many things that fall within this market category and are all purchased and sold through online platforms and middlemen (Market.us, 2024).

The marketing and selling of agricultural products has been completely transformed by the growth of e-commerce. Farmers can sell their products directly to customers through online marketplaces, cutting out middlemen and increasing revenue. Farmers can also create their own e-commerce websites, giving them authority over customer experience, price, and branding. This change gives farmers the ability to access both domestic and foreign markets, in addition to expanding their market reach. Farmers can easily reach a wider audience by avoiding traditional middlemen by utilizing platforms like Amazon and regional agritech marketplaces (Market.us, 2024).

Search Engine Optimization

Agricultural businesses that use SEO tactics are more likely to appear higher in search engine results, which makes it simpler for customers to find them. Visibility is increased by optimizing website content with pertinent agricultural keywords.

Influencer Collaborations

Credibility and reach can be increased by collaborating with food and wellness influencers. Influencers can raise awareness and interest among their followers by sharing real-life product experiences.

Online Auctions

Real-time connections between farmers and purchasers are made possible by platforms that support online auctions and sales of agricultural products, guaranteeing fair prices and more market transparency.

Email Marketing

Email campaigns can be used to distribute instructional materials, seasonal specials, and product updates. Maintaining relationships with current clients and promoting repeat business can be achieved through personalized emails.

Targeted Advertising

Digital marketing makes it possible to provide customized messages to particular client categories. Farmers can reach audiences based on demographics, interests, and habits by utilizing technologies like social media advertising and Google Ads. By guaranteeing that promotional content reaches the people who are most likely to be interested in their products, this precise targeting improves the effectiveness of marketing campaigns and raises conversion rates.

Data Analytics

Data on customer behavior and market trends can be gathered and analyzed due to digital marketing. Farmers can learn which marketing tactics work best, comprehend customer preferences, and adjust their products accordingly by utilizing analytics technologies. Better product creation and more targeted marketing efforts are the results of this data-driven strategy, which facilitates more informed decision-making.

Direct Consumer Engagement

Direct communication between farmers and consumers is made possible by digital marketing, which cultivates a relationship based on loyalty and trust. Farmers may showcase their products, share their stories, and engage with customers in real time on social media sites like Facebook, Instagram, and Twitter. In addition to personalizing the marketing experience, this direct interaction fosters transparency by assisting customers in comprehending the principles and methods underlying their food.

Content Marketing

Another successful tactic in agriculture marketing is content marketing. Blogs, films, and infographics are examples of useful information that farmers and agricultural companies can provide to inform consumers about their goods, farming methods, and sustainability initiatives. In addition to establishing them as authorities in the field, this draws in prospective clients looking for answers and information, which eventually boosts sales.

Cost-effective Marketing Solutions

Digital marketing can often be less expensive than traditional marketing techniques. Numerous platforms and tools for digital marketing provide reasonably priced ways to reach a large audience. For instance, social media may be utilized for very little money, enabling farmers with tight budgets to successfully market their goods and interact with customers.

Developing a Brand Identity

Farmers may create a powerful brand identity by maintaining a consistent online presence and messaging. By emphasizing the distinctive features of their products, like organic farming methods or local sourcing, digital marketing enables the development of a unified brand narrative that connects with customers. A well-known brand can boost market competitiveness and client loyalty.

Mobile Marketing

Mobile marketing techniques, such SMS campaigns and mobile-friendly websites, can successfully engage farmers and customers alike by sending information and promotions straight to their phones, especially in rural areas where mobile usage is strong.

Sustainability Messaging

Marketing materials that emphasize sustainable practices might draw in eco-aware customers. Today's consumers can be effectively reached by sharing stories about environmentally friendly farming practices and the effects of consumer choices.

Barriers to Technological Adoption in Agricultural Marketing

A number of obstacles prevent agricultural technology (AgTech) from being widely adopted and scaled in India. These obstacles include problems with infrastructure, expertise, and farmer confidence in addition to economic difficulties.

Economic and Financial Barriers

Adoption of digital agriculture technologies necessitates large financial outlays for both the initial setup and continuing system maintenance. Farmers have financial obstacles due to the high expenses of technology, such as the acquisition of software and equipment. Furthermore, there is a lack of funding for employee training and maintaining the ongoing use of these technologies. Securing funding for the adoption of technology can also be challenging for farmers, especially small-scale farmers without access to finance.

Problems with Connectivity and Infrastructure

The absence of digital infrastructure, especially in rural areas, is a significant obstacle. The usefulness of digital technology is hampered by the lack of dependable internet connectivity and access to electronic gadgets. Farmers cannot fully profit from digital agriculture if they lack the infrastructure necessary for technologies like cloud computing, sensor technologies, and the IoT to operate at their best (Silva et al., 2023).

Lack of Knowledge and Training

Another major obstacle is farmers' unfamiliarity with digital platforms. This knowledge gap is exacerbated by the lack of training programs and low levels of digital literacy, particularly among older and less educated farmers. Many times, farmers are reluctant to adopt new tools and systems because they don't know how to use technology efficiently. This problem is frequently associated with farmers' age and educational background, which affects their capacity to adopt new technologies (Gaspar et al., 2021; Kaňovská, 2021).

Farmers' Perception and Confidence

Due to a lack of faith in the technologies' potential benefits and applicability, many farmers are reluctant to use digital agriculture technology. They might not think the technology is worth it, particularly if they don't know how it will directly boost profitability, cut expenses, or increase productivity. Furthermore, farmers are discouraged from implementing new technologies due to perceived hazards, such as worries about privacy, data security, and the possibility of more complex farming operations (Gaspar et al., 2021; Kasemi et al., 2022).

Government Regulations and Policy Challenges

Government rules and laws have the power to either promote or impede the use of technology in agricultural marketing. Farmers face obstacles due to imprecise or restrictive rules and insufficient state support for technology integration. Farmers may feel unsupported in their attempts to innovate as a result of the absence of government-backed incentives for technology adoption (Bhattacharyya et al., 2020).

Cultural Resistance

Another significant barrier impeding the use of technology in agriculture is cultural resistance. Due to their strong ties to conventional farming methods, many farmers may be reluctant to adapt. This tension between cutting-edge technology advancements and enduring agricultural customs may cause resistance to implementing new instruments that contradict long-standing methods.

Privacy and Security Issues

Sensitive information regarding farming methods is produced by digital agriculture technologies, which presents privacy and data security issues. Farmers may be reluctant to adopt technology because they fear that their data may be misused or accessed without authorization. To mitigate these worries and foster confidence among farmers, it is imperative to have strong data protection procedures in place (Pauschinger & Klauser, 2022).

External Factors and Risk Perception

The dangers associated with external factors, such illnesses, natural catastrophes, and environmental changes, also affect how technology is used. If farmers believe that these risks are too unpredictable, they may be hesitant to participate in digital agriculture since they are unsure if the technologies will yield a profit. Furthermore, farmers may be discouraged from adopting these technologies due to the perceived difficulty and possibility of failure in integrating them into current systems.

Organizational Type and Farm Size

Technology adoption is also influenced by farm size and organizational structure. While smaller, individual farms may have difficulties because of financial and resource limitations, larger farms and cooperatives typically have greater resources and are more inclined to utilize digital technologies. Technology adoption is also influenced by the kind of farming organization; participation in a cooperative is frequently linked to a higher propensity to use digital tools (Caffaro & Cavallo, 2019).

Other Factors Influencing Technology Adoption

Farmers' decisions to use digital agriculture technology are influenced by a number of factors. These include the farmer's willingness to take chances, the technology's perceived utility, and its usability. Access to training, the availability of technology infrastructure, farm size, financial circumstances, and the farmer's educational background are other crucial elements. Farmers' opinions and choices about adopting technology are also greatly influenced by their family, neighbors, and agricultural advisors (Gaspar et al., 2021).

Overall, digital agriculture has enormous potential in India, but its widespread implementation depends on removing these obstacles. A more effective, productive, and sustainable agricultural marketing system in India will be possible if obstacles related to finances, infrastructure, education, and psychology are removed.

Conclusion

It is vital to address the constraints that may inhibit the adoption of technology in agricultural marketing, despite the fact that the opportunities given by technology in this field are very extensive.

Despite the enormous potential that technology offers in agricultural marketing, it is crucial to remove the obstacles that could prevent its widespread use. To guarantee that all farmers may profit from these developments, factors including farmers' low levels of digital literacy, poor infrastructure, and access to technology must be taken into account. The provision of training, resources, and support systems for farmers requires cooperation between the public and private sectors as well as nongovernmental groups. Unlocking the full potential of technology in agricultural marketing will require initiatives to increase digital literacy, make affordable technology accessible, and provide strong infrastructure.

In conclusion, using technology to promote agricultural products in India is essential to the agricultural sector's future and not just a choice. Farmers may improve their marketing tactics, get beyond conventional obstacles, and eventually raise their standard of living by utilizing technology. It is crucial that all parties involved collaborate as the agricultural landscape changes in order to make sure that technology acts as a driver for progress. In addition to helping farmers, this team effort will support India's agricultural industry's general expansion and sustainability, ensuring food for coming generations and boosting economic resilience in the process.

For the marketing of agricultural products in India, it is crucial to use technology, especially digital marketing and digital platforms for farmers. In addition to improving the success of individual farmers, this integration helps the agricultural industry grow and become more resilient overall, which benefits society as a whole. Adopting technology will be crucial as we negotiate the intricacies of the contemporary agricultural environment in order to achieve sustainable development, food security, and economic empowerment. By prioritizing these advancements, we can ensure a prosperous future for farmers, consumers, and the nation as a whole. The moment has arrived for India to fully embrace the agricultural digital revolution and realize its enormous potential for everyone's benefit. With the right assistance and commitment, the transition to a tech-enabled agricultural marketing framework can result in a flourishing agricultural ecosystem that is advantageous to all.

Acknowledgments

The infrastructural support provided by FORE School of Management, New Delhi, India, in completing this chapter is gratefully acknowledged.

References

AgRevolution. (2024). *DeHaat. Solution for farmers*. https://agrevolution.in/company

AgriApp. (2024). *Our motto*. https://agriapp.com/en/benefits

Ananda, K. R., Sampreetha, H. N., & Chaithanya, G. (2023). Digital agriculture and extension services in quality seed adoption. *Just Agriculture*, *3*(12), 115–122. https://www.justagriculture.in/files/newsletter/2023/august/20.%20Digital%20Agriculture%20and%20Extension%20Services%20in%20Quality%20Seed%20Adoption.pdf

Ascent Group India. (2024). *Mobile apps impacting rural India directly and indirectly*. https://ascentgroupindia.com/blog/mobile-apps-impacting-rural-india-directly-and-indirectly-2/

Bhatt, R., Hossain, A., Majumder, D., Chandra, M. S., Ghimire, R., Faisal, M., Verma, K. K., Riar, A. S., Rajput, V. D., Oliveira, M. W., Nisi, A., Almalki, R. S., Bárek, V., Brestic, M., & Maitra, S. (2024). Prospects of artificial intelligence for the sustainability of sugarcane production in the modern era of climate change: An overview of related global findings. *Journal of Agriculture and Food Research*, 101519. https://doi.org/10.1016/j.jafr.2024.101519

Bhattacharyya, P., Pathak, H., & Pal, S. (2020). Barriers to adaptation of climate-smart agriculture. In *Climate smart agriculture. Green energy and technology* (pp. 155–167). Springer. https://doi.org/10.1007/978-981-15-9132-7_10

Caffaro, F., & Cavallo, E. (2019). The effects of individual variables, farming system characteristics and perceived barriers on actual use of smart farming technologies: Evidence from the Piedmont region, northwestern Italy. *Agriculture*, *9*(5), 111. https://doi.org/10.3390/agriculture9050111

Chhapia, H. (2024). Digitalising Indian agriculture: Impact of major agtech platforms. *Times of India*. http://timesofindia.indiatimes.com/articleshow/109136384.cms?utm_source=contentofinterest&utm_medium=text&utm_campaign=cppst

eNAM. (2024). *e-NAM overview*. https://www.enam.gov.in/web/

FarmERP. (2024). *The changing face of digital agriculture*. https://www.farmerp.com/

Gaspar, P. D., Fernandez, C. M., Soares, V. N., Caldeira, J. M., & Silva, H. (2021). Development of technological capabilities through the internet of things (IoT): Survey of opportunities and barriers for IoT implementation in Portugal's agro-industry. *Applied Sciences*, *11*(8), 3454. https://doi.org/10.3390/app11083454

GoI. (2009). *Brief history of agricultural marketing regulation, its constraints and reforms in the sector*. Government of India. https://dmi.gov.in/Documents/Brief%20History%20of%20Marketing%20Regulation.pdf

GoI. (2013). *Final report of committee of state ministers in charge of agriculture marketing to promote reforms*. Ministry of Agriculture, Department of Agriculture and Co-operation, Government of India. https://dmi.gov.in/Documents/stminprreform.pdf

Haq, Z. (2024). Cabinet clears 7 plans for digital agri push, climate resilience. *Hindustan Times*. https://www.hindustantimes.com/india-news/cabinet-clears-7-plans-for-digital-agri-push-climate-resilience-101725303248726.html

Jain, B., & Dhar, I. (2022). *Digitalisation of agriculture in India*. https://www.investindia.gov.in/team-india-blogs/digitalisation-agriculture-india

Jurgens, J., & Kaushik, P. (2024), *Farmers in India are using AI for agriculture – Here's how they could inspire the world*. World Economic Forum. https://www.weforum.org/agenda/2024/01/how-indias-ai-agriculture-boom-could-inspire-the-world/

Kaňovská, L. (2021). Barriers to and benefits of the use of smart farming technologies for small and medium winemakers, specifically sensors and weather stations: A pilot study. *AGRIS Online Papers in Economics and Informatics*, *13*(1), 71–85. ISSN 1804-1930. https://doi.org/10.7160/aol.2021.130106

Kasemi, R., Lammer, L., & Vincze, M. (2022). The gap between technology and agriculture, barrier identification and potential solution analysis. *IFAC-PapersOnLine*, *55*(39), 314–318. https://doi.org/10.1016/j.ifacol.2022.12.042

Market.us. (2024, March 15). *E-commerce of agricultural products*. https://market.us/report/e-commerce-of-agricultural-products-market/#overview

Pauschinger, D., & Klauser, F. R. (2022). The introduction of digital technologies into agriculture: Space, materiality and the public–private interacting forms of authority and expertise. *Journal of Rural Studies*, *91*, 217–227. https://doi.org/10.1016/j.jrurstud.2021.06.015

Saha, S., Sinha, C., & Saha, S. (2024). Agricultural marketing in India: Challenges, policies and politics. *South Asian Journal of Macroeconomics and Public Finance*, *13*(1), 39–52. https://doi.org/10.1177/22779787231209169

Saraf, S. A., Ali, J., Bahar, F. A., & Mahdi, S. S. (2022). Marketing of agricultural produce in India: Problems and prospects. In F. A. Bahar, M. Anwar Bhat, & S. S. Mahdi (Eds.), *Secondary agriculture* (pp. 85–95). Springer eBooks. https://doi.org/10.1007/978-3-031-09218-3_8

Silva, F. T. D., Baierle, I. C., Correa, R. G. D. F., Sellitto, M. A., Peres, F. A. P., & Kipper, L. M. (2023). Open innovation in agribusiness: Barriers and challenges in the transition to agriculture 4.0. *Sustainability*, *15*(11), 8562. https://doi.org/10.3390/su15118562

Xiong, H., Dalhaus, T., Wang, P., & Huang, J. (2020). Blockchain technology for agriculture: Applications and rationale. *Frontiers in Blockchain*, *3*. https://doi.org/10.3389/fbloc.2020.00007

Chapter 8

Small Farming in the Caribbean Region: Sustainable, Modernised and Technological Alternatives

Andrea K. Veira

The University of the West Indies Cave Hill Campus, Barbados

Abstract

The Caribbean, situated between North and South America in the Tropical Atlantic zone, is home to a diverse range of countries with rich biodiversity and multicultural populations. Many of these islands, classified as Small Island Developing States (SIDS), face significant vulnerabilities due to their small land areas and populations. These islands are particularly exposed to the impacts of climate change, including hurricanes, storms, volcanic eruptions, saltwater intrusion, flooding, droughts and other weather phenomena, all of which threaten biodiversity, food security and economic stability. Notable examples of devastation include hurricanes in Grenada (2004), Barbuda and Dominica (2017) and St Vincent and the Grenadines (2024). Agricultural production in SIDS is typically small scale, relying on seasonal rainfall patterns, with a wet season coinciding with the hurricane season (June to November) and a dry season (December to May) marked by droughts. Crops grown include short-term vegetables like tomatoes and peppers and long-term starches such as sweet potatoes, cassava and yams. Farmers often use traditional methods and government-subsidised inputs but lack evidence-based planning and sustainable practices, making them vulnerable to climate-related disruptions. This chapter explores small farming practices in the Caribbean and how climate change impacts agricultural sustainability. It examines viable, environmentally friendly options for SIDS agriculture, such as agroecology, drones, aquaponics, hydroponics and agrometeorological data. Drawing on examples from Antigua and

Sustainable and Smart Farming Eco-systems:
Changing Landscapes in the Age of Technology, 103–114

Published under exclusive licence by Emerald Publishing Limited
doi:10.1108/978-1-83608-092-320261008

Barbuda, Barbados, St Lucia and St Vincent and the Grenadines, this chapter discusses sustainable solutions that can help maintain biodiversity and food security while adapting to climate change.

Keywords: Sustainable agriculture; Caribbean farming; technology; agroecology; food security

Introduction

Agriculture in the Caribbean presents unique challenges stemming from cultivation practices, climate change and natural disaster effects. This chapter begins with an introduction to the region's geolocation properties and characteristics as well as some of the islands' vulnerabilities. Agricultural practices within the Caribbean are also described generally and using specific country examples, followed by some of the major challenges which affect the agricultural sector within the islands. Proposed solutions to these challenges are explored with a sustainable and more environmentally friendly and safe approach. Emphasis will be placed on a few islands like Antigua and Barbuda, St Lucia and St Vincent and the Grenadines, through the author's research experiences as well as other documented evidence.

The Caribbean Region's Location

The Caribbean region known for its pristine landscapes, warm waters and tropical climate is situated in the Western Hemisphere between North and South America and consists of a chain of islands that are divided into the Bahamas, the Greater Antilles and the Lesser Antilles. It is often referred to as an archipelago of islands within the Caribbean Sea stretching from Cuba in the north to Trinidad in the south (Ahmad, 2011). Some authors refer to the Caribbean as the West Indies, and it is situated such that the Caribbean Sea is on the inside of the island chain/ archipelago while the Atlantic Ocean is on the outside of the chain on the eastern side, with the Gulf of Mexico located nearest to Cuba and the Bahamas. The Greater Antilles consists of Jamaica, Hispaniola, Cuba and Puerto Rico while the Lesser Antilles consists of the other islands in the rest of the chain from the Virgin Islands to Trinidad and Tobago (Encyclopedia, 2019).

The islands range in human population from just over 4,000 persons in the smaller islands to over 11,000,000 persons in some of the larger countries within the Greater Antilles (Worldometer, 2024). The region's geography creates a mixture of volcanic and limestone-based soils in many of the islands which allow for agricultural production to be maximised for tropical crops and animals. Unfortunately, its location also leads to threats as the chain is exposed to the North and Southeast Tradewinds off the Atlantic Ocean, and many tropical storms and hurricanes affect the islands. The weather typically involves high temperatures all year round with a range of 2°C in the south and 5°C in the far north, one dry

(January to April/May) and one wet season (May/June to November/December) each year, with higher heat stress in the wet season and lower heat stress in the dry season (CIMH, Caribbean Regional Climate Center, n.d.). Additionally, earthquakes, volcanic eruptions, droughts and flooding often take place in the region because of several weather systems in the Atlantic Hurricane Season and outside of the season most commonly by trough systems. By geolocation, the entire region is exposed to hazards which can then affect the economy and growth of each island (CCRIF, 2010). These include wind and water hazards which directly impact agriculture, affecting crop and animal production. As a result of the islands' vulnerabilities, and small sizes, among other reasons, some can be classified as SIDS.

SIDS are small developing island states located across the world's oceans that possess very unique characteristics inclusive of geological landforms and processes, heavy impacts of colonisation – often times slavery, and are under threat across every sector because of the hazards presented from the effects of climate change (World Heritage Centre, 2013). In the Caribbean, natural disasters have significantly affected livelihoods across sectors. A few of the more recent ones include the volcanic eruption in St Vincent and the Grenadines in 2021 which also affected neighbouring countries like St Lucia and Barbados, and Category 4 Hurricane Beryl's impact in 2024 which affected both St Vincent and the Grenadines, Grenada and Jamaica. Category 5 Hurricane Irma destroyed the islands of Barbuda and Dominica in 2017. Flood rains triggered landslides which led to major destruction and loss in Antigua in 2019 and in St Vincent and the Grenadines and St Lucia in 2013. Droughts between 2014 and 2016 also affected the Caribbean region. Amid these disasters, the agricultural sectors were severely impacted as the practices within the Caribbean's SIDS remain somewhat antiquated and traditional.

Agricultural Practices in the SIDS of the Caribbean

Agriculture in the Caribbean was historically practised as a monoculture method whereby countries produced one main commodity on a large scale for export and supply to its colonial counterparts in Europe. These included crops like cotton, tobacco and sugarcane. Fruits and spices were also grown and sent back to the mother colonies. Post independence of these islands, the monoculture method continued to take place. For example, bananas were planted in the Windward Islands (Grenada, St Vincent and the Grenadines, St Lucia, and Dominica), a small group of islands within the Lesser Antilles. This was done under a preferential market agreement with England and other European countries. The result was that farmers in these SIDS concentrated on growing many acres of bananas for export as there was a sure market waiting in Europe. Other crops included sugarcane, cotton, tobacco and nutmeg. Many of the crops were incidentally introduced to the Caribbean from colonial ships during colonial times of slavery and later indentureship and remain a large part of the SIDS economic structures. With this diversity of species introduced to the islands, many communities also practised subsistence farming as a form of self-reliance for their families and barter within the communities. Small farming was always very common and remains a large part of the culture within the SIDS of the Caribbean as farmers often own or lease less than an acre of land and grow a variety of roots, herbs, spices, short-term vegetables and fruits and may have one or a few animals (less than 10). The vegetables include tomato, cucumber, carrot, eggplant, lettuce, spinach, cabbage, bell peppers, hot peppers, okra and beans. The animals include goats, sheep, pigs, chicken and cattle. The outputs from this type of farming are often sold in the local vegetable markets by the farmer or by a member of his/her family. The animals are usually slaughtered or sold alive, at key times in the countries calendar of activities, most commonly, when there is high demand often at Christmas, Easter and Carnival or other festivals such as the Creole Festival in St Lucia and Dominica. The supermarkets also purchase fresh produce from the small farmers weekly or as needed.

Another common characteristic of the SIDS is backyard gardening where many persons grow fruits, roots and vegetables in their backyards for home use. This became more common during the COVID-19 pandemic, even though it was very much a cultural practice that perhaps came from the historical subsistence farming practices that existed. It was very common in countries like the Windward Islands where the soil was volcanic and very fertile and arable. Later, during the pandemic, it became a trend and was more common in those islands with other types of soil (e.g. clay).

Another common occurrence within some of the SIDS is family farming. Members of a family work together on creating, maintaining and running a farm inclusive of labour and inputs, and they agree to share the outputs. It is common within rural communities where ancestral/generational knowledge of how to farm was passed from generation to generation. However, in recent times, it appears to end with the pre-millennials as more and more youths are not interested in farming as a career.

One other common feature of agriculture in the Caribbean is the presence of farming cooperatives which form to support farmers in technical and financial ways where possible. Farmers who join the cooperatives are often able to access financial funds through loans, aid or otherwise and are privy to training and assistance with technical knowledge that can assist their farm operation and production. This is common in the French Caribbean islands as well as others like St Lucia. In many cases, the suggestions and plans are to pool together lands for specific commodities which allow for the small farmers to work together to grow a specific crop for provision to the markets. This allows greater pooled yields. However, the cooperatives do not function very effectively in the Caribbean as many farmers do not honour agreements and contracts and seldom pool to help each other.

Farmer Field Practices in the SIDS of the Caribbean

Farming practices in the SIDS typically involve extensive farming whereby the farmer uses very little inputs. Often, the cost of inputs is too high and so the farmer is unable to purchase many inputs to shape an intensive system for the small land sizes used in small farming. Some farmers use synthetic fertilisers and herbicides and pesticides and, in many cases, utilise donated or government-subsidised inputs without soil analysis and technical expertise on the farming system. Some farmers use ancestral/generational farming knowledge passed on from their families and utilise animal manure and vegetable/green waste to fertilise the soil. Plants with beneficial properties like pest repellence or animal feed source may also be planted close to the crops or used around the borders of the farm if ancestral knowledge indicated that they can serve as an insecticide or other useful purpose such as wind breaks/barriers and animal food. Manual clearing of the land and/or burning the land with fire and in some instances chemicals which are often restricted in the developed world are also used by the farmers and with little personal protective gear. In a few instances, some machines are used for land clearing, preparation and harvesting. However, this is not very common and tends to occur only on the flat areas with little to no knowledge of the effects of tillage. On the slopes, the common practice is terracing that follows the land's natural contours and the formation of beds and or banks that are constructed manually by the farmer. In many instances, small farming in the Caribbean is very labour intensive. Farmers also plant based on the season as in the Caribbean there is a wet season and a dry season as mentioned earlier. Crops are planted strategically to be able to capitalise on the rainfall during the wet season as irrigation systems are not widespread. A few farmers use drip or sprinkler irrigation systems from rivers, ponds or the national water systems but many rely heavily on rainfall and only harvest rainwater in 55-Gallon Blue Drums placed on farms. These limit production in dry conditions. Most production systems are open field while a few are in tunnel houses with irrigation systems and shade cloth. The cultivated crops in these houses tend to be short-term crops like cucumber, beans, tomato and bell peppers. In the small farming community, pests and diseases are commonly treated with synthetic chemicals available on the local market. In a few instances, pests are manually removed when the attack is minor. Other farmers suffer losses as they are unable to treat chemically and do not utilise any other methods.

A very common feature of small farming production in some of the Caribbean islands is the lack of crop scheduling and planning as well as lack of record keeping. This significantly impacts the availability of crops for the markets and leads to increased imports as many roots and vegetables are seasonal. The market is often either flooded with a particular vegetable or that vegetable is scarce and so the market prices are affected. In St Lucia, in 2019–2020, over 100 farmers were involved in a Caribbean Community (CARICOM) – Japan Overseas Friendship Project executed by the Caribbean Agricultural Research and Development Institute (CARDI) which gave access to a crop scheduling application on their devices and training in understanding the market surpluses and demands to plan and schedule cultivation of crops. However, scheduling is not the normal practice and continuity of the use of the app as well as expansion of training and support did not extend beyond the project. All these farming practices are unfortunately not able to survive the effects of climate change that now surrounds our daily lives.

Climate Change Effects on Agriculture

As many farmers continue to use the practices described in the previous section, the effects of climate change on many of the islands within the Caribbean, especially the small islands, have already surfaced and continue to present challenges to the farming community. These include droughts, floods, landslides, saltwater intrusion, desertification of the land and ineffective pollinators. Additionally, there is increased soil acidification, the need for increased water conservation and usage, and a perception that increased synthetic fertiliser use is necessary. All of these effects stem from changes in climate over the decades which are now showing up even more prominently in the smaller land spaces of the very vulnerable SIDS. The climate affects the temperature, humidity and water levels which in turn affect the soil chemistry and physics and hence growth of plants as well as livelihoods of animals. The chemical ions within the soil and its properties are also affected which affects the plants growth and eventually filters into the food chains affecting animals like large and small ruminants and eventually humans.

Droughts result in extreme dry conditions and water shortages which means that small farmers who rely on rainfall and who harvest only minimal amounts using low technologies are not able to obtain sufficient water to grow crops or hydrate and care for their animals. Contrastingly, floods result in too much water that cannot drain quickly enough from farmers' fields and end up waterlogging and killing plant roots. The balance of ions within the soil is also affected as many nutrients are leached or washed away. Similarly, any synthetic inputs applied by the farmer such as insecticides and fungicides may be wasted as they are washed off during floods and can enter ground water systems. Excessive rains and floods can also lead to land slippage through landslides whereby land, crops grown on the lands and animals can be destroyed during the slippage.

Climate change also causes saltwater intrusion. This is known to be taking place in different areas of countries within the Caribbean such as the Marquis Estate in St Lucia. This takes place when saltwater moves inwards towards the land along coastal areas or when it infiltrates freshwater aquifers which results in increased soil salinity affecting the soil chemistry and uptake of nutrients by the plants (USDA, n.d.). It is a direct result of climate change and rising sea levels which result in higher tides, increased storm surge and raising of the groundwater table. Farmers therefore suffer loss as many crops die from high salt conditions and farmers often lack the knowledge on mitigating measures that can be used to reduce the problem. This problem is further compounded during extremely dry conditions or droughts as there is little rainfall to wash the salts out of the soil through leaching.

Another challenge is desertification of the land. In the Caribbean, increased temperatures and water scarcity as a result of climate change, along with unsustainable agricultural practices are resulting in soil loss, degradation of habitats, reduced biodiversity and reduced ability to withstand floods and droughts (FAO, 2024a). The land is degraded and becomes difficult to cultivate crops successfully. It also becomes harder to obtain feed for animals. The SIDS like Antigua and

Barbuda, St Kitts and Nevis and other islands within the Leeward Islands of the Caribbean are more at risk of this challenge as they are water-scarce countries. It is very important that farmers are trained and educated on how to prevent this process and how to efficiently use and conserve water. Unfortunately, many farmers do not practise any type of water conservation or efficient water use except for a few who utilise covered dams/ponds in the dry season and drip irrigation. Traditional methods and perceptions still dominate where some farmers think that sprinkler irrigation systems are more efficient than drip systems and many do not put adequate water harvesting systems or conservation methods in place to take care of their crops and animals during the dry season. There is also a perception by many farmers that synthetic fertiliser can solve many of their problems. This results in fertiliser being used without proper soil testing or analyses and often leads to soil acidification or alkalinisation which affects the balance within the soil chemistry and the plant's uptake of nutrients. This overuse and sometimes incorrect use of chemicals affects the soil biology and chemistry. Other synthetic inputs commonly used like insecticides, herbicides and fungicides can also be detrimental to the pollinators.

Pollinators in agriculture are vital as without the bees and other pollinating insects, there would be no pollination to lead to reproduction and food production. Unfortunately, many of the synthetic chemical inputs used by farmers kill the pollinators. Additionally, with increasing temperatures because of climate change, bees are not able to work and function optimally to pollinate the food crops. Bee foraging and nesting can be limited and reduced if there is reduced flowering from plants that are under heat stress (Walters et al., 2022). Additionally, heavy rainfall leads to flooding and can reduce foraging time for bees as well as displace their hives and habitats (Durant, 2022). The impacts of the changes in climate affect the bees' functionality which then affects plant production. Unfortunately, with climate change also affecting pests on crops, farmers may use more pesticides which are also harmful to the bees. With less pollination taking place as the bees' health are compromised, then the amount of available food from crops also declines and threatens the food security of the islands. It is therefore imperative that sustainable approaches to agriculture are used as solutions to many of these problems.

Sustainable Solutions

Sustainable agriculture involves safeguarding the environment through biological use of products to help with food production and food security (Shelef et al., 2018). It involves land and resource management and human and ecosystem interface so that food is produced along with its byproducts while taking care of all resources involved. It represents a system that small farmers in the Caribbean's SIDS should adopt to preserve and conserve its most treasured assets – land, environment, biodiversity and humans. Agricultural methods, techniques and technologies which can help build a sustainable agriculture sector and present solutions to the problems experienced because of climate change include agroecology, technology use, research, aquaponics, hydroponics, controlled systems, renewable energy systems, smart irrigation techniques, biological pest control and integrated pest management (IPM). Each of these will be discussed from a Caribbean SIDS, small farming perspective.

Agroecology involves practising agriculture with an environmental and ecological consciousness that ensures the preservation and not the destruction of our environments. The environments in this context mean the land, the humans farming the lands, the biodiversity and all other biotic and abiotic factors that allow the land to produce. It is a practice that utilises sustainable methods, appropriate technologies, traditional knowledge of farmers and does not significantly affect the environment (Alves & Neneve, 2019). Many of the agroecology approaches like cover crops, green manures, intercropping, agroforestry and crop-livestock interactions serve many purposes as they lead to favourable changes in the entire farming system (Gliessman, 2000). These techniques are common in the non-English-speaking Caribbean countries like Cuba and Guadeloupe. In St Vincent and the Grenadines, hedges and mulches of beneficial plants like *Gliricidia sepium* (local name Glory Cedar) and *Azadirachta indica* (local name Neem), which are otherwise considered to be plants without any significant use, were used in an experiment to grow two varieties of sweet potatoes. The purpose was to introduce an agroecological approach that adds nutrients, repels pests and is safe and sustainable. Overall, one of the aims was to reduce the amount of synthetic chemicals used in the farming of roots using sweet potato as the example. The results were promising as the mulch increased the total marketable yields of the sweet potatoes, and closer to the hedge, there was less *Phyllophaga* spp. (grub) damage (Veira, 2023). This study also included a human component to determine the farming practices of the sweet potato farmers in the sample, as well as their perceptions towards various methods in agriculture inclusive of synthetic chemical use and environmentally safe practices. This was included in the agroecological study as the importance of the humans and of capturing farmers' practices and traditional knowledge was recognised as a component of any form of agroecology (FAO, n.d.). Results from that study captured a traditional technique of using spoiled fish *Decapterus macarellus* commonly called 'Robin' to kill rats in the field. This type of knowledge passed on from generation to generation, along with evidence that agroecological methods can work and will not affect yields negatively is very useful. Other agroecological approaches can help to add nutrients and replenish exhausted soils, while protecting and encouraging the biodiversity of the environment and micro-environments as well as manage saltwater intrusion. Approaches such as utilising forest trees as wind break barriers and soil stabilisers, water attractant forestry techniques and salt tolerant plants used through agroforestry, creating whole systems of animal and crop interactions whereby the animal manures fertilise the crops, and the crops feed some of the animals, along with the use of cover crops from the legume family are all useful. These practices can mitigate against some of the climate change factors affecting agriculture in small farming. They can also protect the pollinators and the soil composition and structure as well as reduce and possibly even eliminate the heavy and sometimes unnecessary use of synthetic chemicals. Agroecology allows for more efficient water use when needed by the plants and conservation during dry conditions. This option of sustainable agriculture can even be further improved with various technologies.

Technologies that can complement agroecology include smart irrigation systems, drones, biological pest control, pheromone traps, IPM, improved seed varieties, crop scheduling mobile applications (apps) and agri-climate prediction software. Smart irrigation technologies with built in training on their efficient use, along with tensiometers and soil moisture probes will allow farmers to understand water needs and use for their crops. This means little wastage of water through technologically controlled drip and fertigation systems. Another technique of buried irrigation systems can also conserve and more efficiently allow the water to be utilised by the plants.

Water use, pests and diseases can be monitored using drones. These devices can determine disease and pest infestation in fields so that supporting and appropriate reduced synthetic chemical IPM routines can be established. In such routines, a safer method of biological control that uses beneficial microorganisms or natural predators as well as pheromone traps can be researched and developed to treat and alleviate pests and disease problems. This should also include generational farmers' knowledge to capture culturally relevant and accurate farmer practices. For open field production, research on heat, drought and pest tolerant seed varieties at the farmer level and through farmer field schools (FAO, 2024b) can transform the quality and shelf life of agricultural outputs. Crop scheduling apps, built to each country's market demands and supplies can also encourage farmers to plant smartly and to gain good value for their produce. An example of such an app was introduced to a small group of farmers in St Lucia through a project executed by CARDI (Veira, 2021), but only a few farmers continued to use it after the project ended. Unfortunately, farmers tend to work in isolation, and this requires a cultural change that can perhaps be integrated into farming communities slowly through working cooperatives. This would require continuous training and the use of profit examples to allow small farmers to start working together on crop schedules, inputs and outputs to make the industry more profitable. Capturing agrometeorological data using it in crop simulation modelling software can allow for further research to be done on production systems with climate changes. This will serve to enhance production and output. A combined approach that looks at the entire agricultural system holistically and not in isolation can be very beneficial. This includes energy sources on farms. Renewable smart sources like solar or wind can fuel the technologies described earlier in addition to any climate smart agricultural systems, water catchment systems and value-added processing facilities.

Establishing renewable power sites in communities for farmers can allow them to create closed systems for agriculture such as smart greenhouses with cooling and irrigation systems as well as alternative modernised methods such as aquaponics and hydroponics. These systems will transition traditional farming to modernised systems where loss can be minimal if the systems are utilised effectively. This would include efficient water use, recycling of nutrients within the systems, use of high producing varieties and disease and pest control through agroecological and/or organic methods. This was successfully done with three young farmers in St Lucia as they produced heirloom bell peppers, tomatoes, lettuce and tilapia for the hotel and food industry weekly through their respective smart greenhouse, hydroponic and aquaponic system. The use of technology and modernised methods for farming and agriculture can become widespread in the Caribbean and can serve as the magnet for the youth to get actively involved in agriculture. A rebranding of agriculture as a food security business for young entrepreneurs with clear profit gains from low intensive labour will be more attractive to the sector.

Conclusions and Reflections

Agriculture in the Caribbean has been heavily influenced by colonial monoculture systems, subsistence and small farming as well as generational traditional methods and knowledge. As a chain of small islands, small farming communities often suffer the impact of climate change and modifications and adjustments of practice towards a more modernised and sustainable practice is not widespread. Sustainable agriculture should for Caribbean small farming communities involve agroecology inclusive of agroforestry techniques, current technologies, smart closed agricultural systems and use of renewable energy in farm and value-added operations. However, as many of the islands of the Caribbean are SIDS several key themes and areas must be carefully intertwined to allow such an important change in practice.

One key theme is continuous education and training of farmers in modernised sustainable agricultural methods at a level that is encouraging and motivating to the farmers. This can include a new extension support framework that focuses on modern technologies and methods of agriculture as well as training through modern delivery modes both for extension officers and farmers. This type of training can also encourage youth in farming. Advocacy for youth to be involved in agriculture must involve reshaping of the sector and career as a smart technology field that is less labour intensive but more efficient and practical through technology integration. This technology can then be incorporated with sustainable agriculture that is safer for the environment, the farmer and the consumer.

Sustainable agricultural approaches as another key theme when shared through success stories and an economic benefit showcase can lead to cultural adjustments and changes in farming. A lead farmer approach can be incorporated whereby the farmer who is successful at this approach shares and encourages other farmers. Throughout this approach, continuous research in all areas of sustainable agriculture to provide farmers with evidence-based examples of its success is necessary. The use of evidence-based relatable case studies and promotion of sustainable agricultural practices through real-life examples can allow the farmer to visualise and adopt a successful path towards sustainability in farming. One area that needs to be guided and available to farmers for sustainable agriculture to take place is the next theme of financing.

Financing that is attractive and affordable with numerous opportunities to modernise the sector must be available to the small farmer and to various farmer groups. Financial and marketing training for farmers is also very important, so that they can see the economic benefits of sustainable practices and conduct their farming as a business/entrepreneurship platform. This type of training can be afforded through improved partnerships between stakeholders and farmers leading to a stronger understanding of food security from the financial business perspective. Financing opportunities are directly linked to policies and legal agreements which must be present to support an environmentally safe alternative to agriculture inclusive of the use of modern technologies and even the functionality of cooperatives. If policies are in place to support the small farmers on these modernised adjustments and transformations to agricultural systems in the Caribbean region, agriculture can become a more attractive and sustainable sector.

As a region of incredible natural beauty and top tourist attraction sites, it is imperative that the sector transforms and makes much needed changes that allow for the protection and safeguarding of our lands, our people, our flora, our fauna and our food. Through sustainable practices and actions that lead to accomplishing a holistic view of all systems involved in food production, we can achieve food sovereignty and food security if so desired. The question is, 'Is this what we want and are we prepared to make it a reality?'

Implications for Policy

If this is indeed what we as a Caribbean wish to make a reality, there are several implications that exist for policy. These include but are not limited to environmental, social, economic, educational and ethical. Environmentally, we must ensure that policies are written to protect our people, our lands and our species diversity from any technologies that are implemented in agriculture. A low carbon footprint should be present with these technologies to ensure that their use is not contributing to climate change. Preservation and conservation of endemic species of our SIDS must also be in place, as well as protection from any technologies that are used to assist in maintaining our food security as a region. Socially, policies must address different income brackets of consumers and producers to ensure that all persons are food secure. This can be done through access to better nutrition through sustainably produced foods which can then lead to healthier lifestyles with the potential for reduction in the non-communicable diseases that plague our region. Economically, policies that allow for equitable access by all levels of farmers (all income brackets) to loans, subsidies, incentives, financing, business start-ups and otherwise must be in place so that fairness to access opportunity exists. Educationally, mandatory training within educational institutions from primary through to tertiary levels inclusive of adult education units should be instituted to allow knowledge transfer in the modernised agricultural methods, technologies and approaches. Ethically, policies must also be written to protect the ownership of the data that come from the use of technologies as well as the unique genetic materials that may be present as value-added products in agriculture. If these implications are appropriately addressed by policymakers and supporting institutions, small farming in the Caribbean can transform in such a way that our region achieves food security in a safe manner.

References

Ahmad, N. (2011). *Soils of the Caribbean*. Ian Randle Publishers.

Alves, A., & Neneve, M. (2019). Agroecology, the interaction between agriculture and environment: An example from Cuba. *International Journal of Advanced Engineering Research and Science*, *6*(10), 216–221. https://dx.doi.org/10.22161/ijaers.610.33

CCRIF. (2010). *Enhancing the climate risk and adaptation fact base for the Caribbean – Preliminary results of the ECA study*. Grand Cayman.

CIMH, Caribbean Regional Climate Center. (n.d.). *Characteristics of Caribbean climate*. https://rcc.cimh.edu.bb/caribbean-climatology/#:~:text=The%20climate%20in%20the%20Caribbean,November%2030th%20of%20each%20year

Durant, J. (2022). *Bees face many challenges – And climate change is ratcheting up the pressure*. https://theconversation.com/bees-face-many-challenges-and-climate-change-is-ratcheting-up-the-pressure-190296

Encyclopedia, N. W. (2019). *Caribbean*. https://www.newworldencyclopedia.org/p/index.php?title=Caribbean&oldid=1029350

FAO. (2024a). *Action against desertification*. https://www.fao.org/in-action/action-against-desertification/countries/caribbean/en/

FAO. (2024b). *Farmer field school approach*. https://www.fao.org/agriculture/ippm/programme/ffs-approach/en/

FAO. (n.d.). *The 10 elements of agroecology guiding the transition to sustainable food and agricultural systems*. http://www.fao.org/3/I9037EN/i9037en.pdf

Gliessman, S. R. (2000). *Agroecology: Ecological processes in sustainable agriculture* (E. Engles Ed.). Lewis Publishers.

Shelef, O., Fernández-Bayo, J. D., Sher, Y., Ancona, V., Slinn, H., & Achmon, Y. (2018). 2 – Elucidating local food production to identify the principles and challenges of sustainable agriculture. In C. M. Galanakis (Ed.), *Sustainable food systems from agriculture to industry* (pp. 47–81). Academic Press.

USDA. (n.d.). *Saltwater intrusion: A growing threat to coastal agriculture*. https://www.climatehubs.usda.gov/hubs/northeast/topic/saltwater-intrusion-growing-threat-coastal-agriculture

Veira, A. K. (2021). Action point 5: Implementing practices at field level. Saint Lucia: Advancing the use of information and communication technology solutions for climate-smart agricultural practices. In *Climate smart agriculture case studies 2021: Projects from around the world* (pp. 77–79). FAO. https://doi.org/10.4060/cb5359en

Veira, A. K. (2023). *Agroecological approaches to sweet potato farming in the Caribbean* [PhD, Biology doctoral thesis, Unpublished]. The University of the West Indies, Bridgetown, Barbados.

Walters, J., Zavalnitskaya, J., Isaacs, R., & Szendrei, Z. (2022). Heat of the moment: Extreme heat poses a risk to bee-plant interactions and crop yields. *Current Opinion in Insect Science*, *52*, 100927. https://doi.org/10.1016/j.cois.2022.100927

World Heritage Centre. (2013, February 2013). Small island developing states. *World Heritage*, *66*.https://whc.unesco.org/en/review/66/

Worldometer. (2024). *Caribbean population*. https://www.worldometers.info/world-population/caribbean-population/.

Chapter 9

Past, Present and Future of Indian Smart Farming Ecosystems: A Review

Arushi Bathla[a], Chetna Chauhan[b] and Ashish Gupta[c]

[a]*School of Liberal Arts and Humanities, O P Jindal Global University, Sonipat, Haryana, India*
[b]*Universidad de Los Andes, Bogota, Colombia*
[c]*Faculty of Management, South Asian University (SAU, an International University Established by SAARC Nations), New Delhi, India*

Abstract

This chapter synthesizes extant literature and identifies the current intellectual frontiers in the field of smart farming ecosystems in the context of sustainability and technology in India. We employed stringent protocols through systematic literature review (SLR) to select relevant articles and then we undertook a thematic analysis using bibliometric analysis. Finally, we employed a barriers, mitigation strategies and opportunities (BMO) framework to present the status quo of the field. At first, we identify and explain various clusters through bibliometric coupling. Second, through co-word analysis, we identify and explain additional clusters. Finally, through the BMO framework, we elaborate on the barriers in the process of successful implementation of smart farming in India, mitigation strategies and opportunities such as investment in sensor technologies for monitoring the real-time processing activities, among others. This study conducted a comprehensive review of the field, which to our limited knowledge no prior studies have done so far. Besides, the study also outlines interesting research questions for future researches. For investigating many contemporary challenges related to smart farming ecosystems, like climate sensitivity, data security and privacy issues, lack of farmer unity and empowerment, etc. have been outlined. This study serves as a starting

Sustainable and Smart Farming Eco-systems:
Changing Landscapes in the Age of Technology, 115–127

doi:10.1108/978-1-83608-092-320261009

point for academicians and policymakers in assessing and evaluating the impact of long-term initiatives on smart farming, assisting in identifying areas for improvement and allocating resources accordingly.

Keywords: Smart farming ecosystems; sustainability; technology; systematic review; framework

1. Introduction

Agriculture plays an important role in sustaining all human activities. However, complex challenges such as overpopulation, lack of resources poses, food insecurity, among others demand continuous advancements such as smart farming (Harisudin et al., 2023; Islam et al., 2021; Ouafiq et al., 2022; Ragazou et al., 2022; Yadav et al., 2022). Today, agricultural systems require a strategic shift from the current paradigm of mere productivity to sustainability and smart farming. Sustainable practices not only focus on enriching agricultural productivity but also help to reduce harmful environmental impacts.

The digital technologies such as Internet of Things (IoT), mobile technologies and devices, data analytics, artificial intelligence (AI), digitally delivered services and other applications are influencing the decision-making (Orjuela-Garzon et al., 2021; Osrof et al., 2023; Quintero et al., 2022; Tama et al., 2023). Numerous examples demonstrate the use of digital technologies at different stages of farming activities such as automation of farm machinery resulting in reduced labour input; use of sensors and remote satellite data for improved monitoring of crops, land and water; and IoT for agriculture product traceability (Akbar et al., 2024; Balkrishna et al., 2023; Ennouri & Kallel, 2019; Javaid et al., 2022).

In an Indian context, agriculture is the backbone of the Indian developing economy, yet smart farming technologies are presently at the nascent stage. Although various initiatives such as precision farming projects, e-choupals, laser land levelling (Eastwood et al., 2019; Jakku et al., 2019; Knierim et al., 2019), have been employed to provide weather information, disease forecasting, expert crop advice systems to increase yield and efficiency, yet their application face complex challenges due to poverty and landholding size. To make a significant contribution to Indian agriculture systems, a thorough analysis of the current status of art literature and practices in the field of smart farming and sustainability is crucial (Atanasov, 2023; Pero et al., 2024).

Although, in the past, a few review studies were conducted on sustainability and smart farming applications. These studies have focused on niche such as machine learning (ML) or artificial learning applications or use of block chain (Orjuela-Garzon et al., 2021; Osrof et al., 2023; Quintero et al., 2022; Tama et al., 2023), covering specific aspects like supply chain risk management (Eastwood et al., 2019; Jakku et al., 2019; Knierim et al., 2019) or sectors (Quintero et al., 2022) or crop based (Atanasov, 2023; Pero et al., 2024).

However, to the best of our knowledge, not much attention has been paid on emerging economy perspective; in a country-specific level in the Indian context which have conducted a comprehensive review the past, present and future prospects of smart farming ecosystems in the context of sustainability and technology.

Therefore, in this study, we present an SLR (Bathla, Chawla, & Gupta, 2024) of 47 papers. It is anticipated that the agricultural sector will continue to see the increasing adoption of smart technologies in India in the future. The results of this study will guide the researchers and practitioners to understand the present status of these applications. This will help them to understand how adoption of smart farming technology can support the farmers and the economy to optimize farming practices to increase yields, crop quality and incomes in a sustainable manner.

Based on this backdrop, this study aims to address the following research questions:

RQ1. What is the intellectual structure of literature in smart farming ecosystems and sustainability in India?

Objective 1. To undertake a bibliometric analysis and content analysis to outline the thematic structure of the field.

RQ2. What holds in the future for this field?

Objective 2. To present a framework of the field to trace the directions for future research.

The remainder of this chapter is structured as follows: methodology, findings, discussion, implications, conclusion and limitations.

2. Methodology

SLRs are known to involve a narrow focus since the former depends on qualitative approaches and hence may be influenced by interpretation bias from researchers, especially if scientific quality evaluations of selections are not conducted. On the other hand, bibliometric analysis relies on quantitative approaches, which may help limit such bias and thus can be additionally conducted to examine the structural connections amid research constituents within a field. However, bibliometric analysis is descriptive in nature, and therefore, to overcome this limitation, an in-depth critical content and text analysis is required. Content and text analysis identify insights in the text to draw interpretations for the empirical starting point of generating research scope. For this chapter, we adopted an SLR using Preferred Reporting Items for Systematic Reviews and Meta-analysis statement framework, along with both bibliometric content analyses to cater to *RQ1*, while text analysis was conducted to cater to *RQ2*.

2.1. Phase 1 – SLR

SLRs are widely accepted to help academics synthesize a niche body of literature. They are also believed to be more appropriate for specialist study fields, smart farming and sustainability, for instance (in the context of this study), which in itself is an emerging field with a few publications.

2.1.1. Scope

The scope included determining the keywords, article selection criteria, database, quality evaluation methods and descriptive statistics (top journals and top authors) of the field.

2.1.2. Protocol

The protocol of this section encompassed a selection of Scopus database, which is in line with the prior SLRs attempted in the arena of sustainability, technology, and smart farming ecosystems.

2.1.3. Database Curation

For the data-curation process, we constituted a panel inclusive of the authors, along with one design and management expert each (excluding authors) to review some of the seminal papers on sustainability, technology, and smart farming and determine appropriate search strings that effective yielded in two search terms. We executed the search with recommended search terms, first; 'sustainability' AND 'smart farming' AND 'technology' AND 'India', second: 'sustainability' AND 'smart farming *ecosystem*', AND 'technology' AND 'India'. The second search string was later excluded, since it had an extremely narrow scope and yielded only a few results. Each of the authors independently used the selected search string on the Scopus database (Bathla, Chawla, Hofaidhllaoui, et al., 2024). See Fig. 9.1 for data curation. After examining the articles according to article selection, criteria A and B1, a robust manual screening/rating-based staged review by the authors was independently undertaken for extracting relevant papers after reading all the abstracts in depth, taking article selection criterion B2 as the basis. This resulted in 47 relevant documents for the study; 24 articles were summarily removed during this process. The ratings were scrutinized for inter-rater reliability employing the Kappa statistics (0.79), signifying that the procedure had a strong/appropriate/substantial level of agreement.

2.2. Phase 2 – Bibliometrics

Technological innovation today has aided in executing this bibliometric analysis, whereby we could closely examine the structural constituents of a research field, enabling us to better interpret and understand data meaningfully (Yadav et al., 2022).

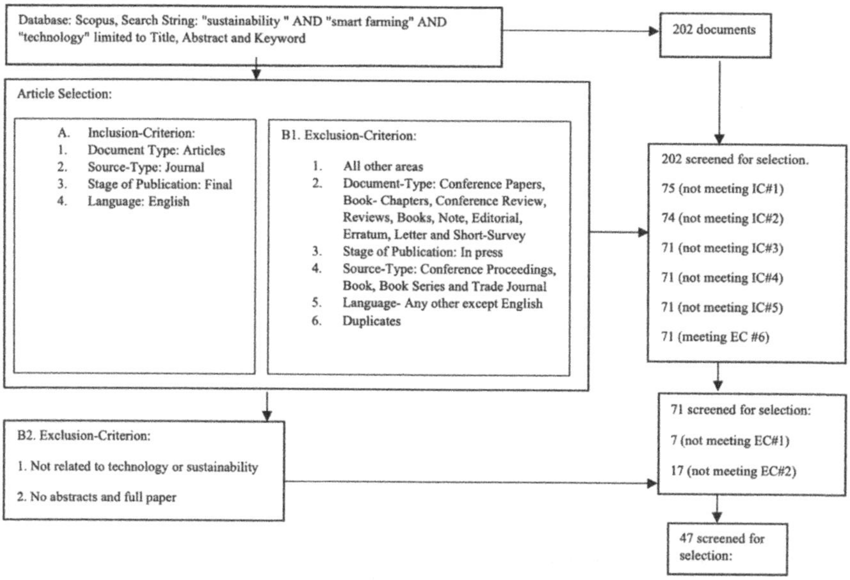

Fig. 9.1. Data-curation Process. *Source*: Authors' own.

2.2.1. Scope

We limited the scope by using only 47 articles/documents for analysis.

2.2.2. Protocol

The protocol for this review included using the VOSviewer software, which is a tool for constructing and visualizing bibliometric networks. We applied association strength normalization, fractional counting and default values for visualization in VOSviewer.

2.2.3. Science Mapping Analysis

The essence of visualizing bibliometric networks, often known as science mapping, has garnered considerable interest, because it is an efficient method for investigating the relationship of interconnected networks.

2.2.3.1. Bibliometric Coupling. Multiple common references amid two or more publications indicate deeper relation in the subject matter dealt with in such studies. The unit of analysis for coupling was articles/documents. We undertook a threshold of two citations for each document, and subsequently, we got 31 relevant documents out of 47. Further, out of these 31 documents, only 19 consisted of the large set, and hence, only they were used for bibliographic coupling analysis. From this large set of 19 documents, we finally got 5 themes/clusters (Fig. 9.2) on network visualization. As an acceptable practice, the top-cited articles

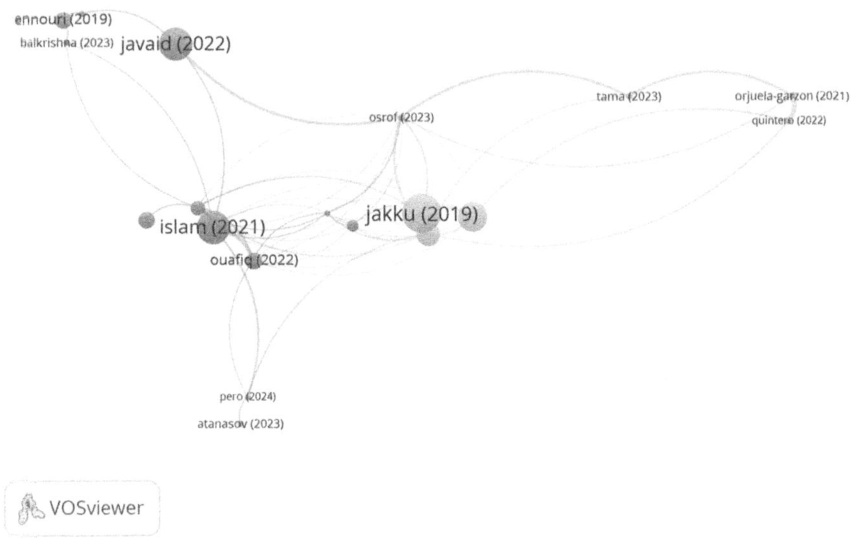

Fig. 9.2. Bibliometric Coupling. *Source*: VOSviewer.

under each theme/cluster were identified as shown in Table 9.1. Then, we analysed each theme to further identify the dominant themes and sub-themes as shown in Table 9.2.

Next, we employed content analysis on the 47 articles/documents extracted above to better understand the themes' insights and confirm the robustness of research/bibliometric coupling findings. We selected articles with 25 and more citations for content analysis. We found that using both bibliometric coupling and content analysis, five themes emerged, confirming thereby the legitimacy and rigour of the surfaced themes/clusters that are described in detail below:

Table 9.1. Top Papers of Each Cluster.

Cluster	Papers
1	Harisudin et al. (2023); Islam et al. (2021); Ouafiq et al. (2022); Ragazou et al. (2022); Yadav et al. (2022)
2	Orjuela-Garzon et al. (2021); Osrof et al. (2023); Quintero et al. (2022); Tama et al. (2023)
3	Akbar et al. (2024); Balkrishna et al. (2023); Ennouri and Kallel (2019); Javaid et al. (2022)
4	Eastwood et al. (2019); Jakku et al. (2019); Knierim et al. (2019)
5	Atanasov (2023); Pero et al. (2024)

Source: Authors' compilation.

Table 9.2. Themes and Sub-themes of Each Cluster.

Main Theme	Sub-theme
Devices	Unmanned aerial vehicles, low-power consumption-embedded devices, unmanned vehicles, environmental performance, profitability, central Java
Programmes and models	Massive technology adoption programme, technology transfer, agent-based modelling
AI	Deep learning-enabled computer vision techniques, NodeMCU, greenhouses
Farmer perspectives	Farm advisory services, trust, transparency and benefit sharing
Technology	Remote sensing, viticulture optimization

Source: Authors' compilation.

Theme 1: Research in this field is focused on various devices used for minimizing costs, resource efficiency, and gaining robustness in the transmission of the data to the cloud with edge computing, long range wide area network, low-power consumption-embedded devices, IoT and wireless sensor networks, guaranteeing the system's durability and the data modelling among others.

Theme 2: Research in this field is focused on programmes and models that increase crop competitiveness and sustainability by implementing a technology transfer programme, agent-based modelling paradigm. Research has proposed models that make it possible to study the dynamics of local learning and how patterns of specialization emerge and to improve the transfer and adoption of technologies (smart farming), increasing their productivity and sustainability and reducing their environmental impact in an agricultural innovation system.

Theme 3: In this area, technologies are being applied within greenhouses for growth monitoring, disease detection, yield estimation, crop yield, pest control, monitor temperature, soil health/moisture, etc. in real time. The tools used are based on data mining such as remote sensing and NodeMCU-linked sensors and imagery technologies, such as picture treatment and analysis.

Theme 4: The findings in this field show that smart technologies exhibit potentially disruptive features for farm management, necessitating greater input from a farmer's advisory network to facilitate optimal farm system adaptation. The analysis shows that issues related to trust are central concerns for many farmers. Evidence indicates that farmers' behaviour regarding the adoption and continuation of smart framing is affected by their socioeconomic and psychological factors. These include procedural concerns about transparency and distributional concerns about who will benefit from access to and use of 'farmers' data'. These concerns create scepticism about the value of 'smart' technologies among some industry stakeholders, especially farmers.

Theme 5: The area studies the feasibility and efficacy of the ML models, demonstrating their potential to revolutionize vineyard management and contribute to sustainable agricultural practices.

2.2.3.2. Co-word/Co-occurrence Analysis. When words emerge often across texts, it signifies that ideas underlying those words are inextricably linked with the same thought. This technique is of critical importance, as it constructs a similarity score based on the actual content of the documents. This semantic map aids in understanding its cognitive structure. A set of maps created over time periods can be used to track the development of the conceptual space and could help academics better cognize the thought process and progress in the field.

The words across all 47 documents were extracted from document title, keywords and abstract and formed the unit of analysis. The minimum no. of occurrences of a term was kept at three. This resulted in 1,938 terms, out of which, 137 terms met the threshold of 3 occurrences. Sixty percentage of these 137 terms (i.e. 82 terms) were most relevant. As may be seen from Fig. 9.3, overlay visualization for co-occurrence analysis of text data, IoT, innovation, remote sensing, irrigation, greenhouse gas emission, robotic, AI, emerged among the most relevant words. Furthermore, the most recent items under study were ML, fertilization, food productivity and farming practice. These networks were studied to find similarities to categorize into themes and their respective items as shown in Table 9.3.

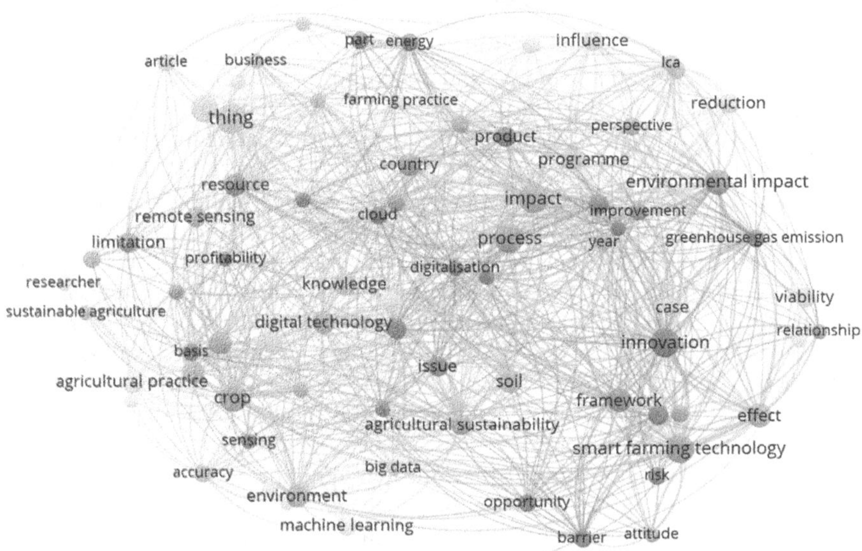

Fig. 9.3. Overlay Visualization for Co-occurrence Analysis of Text Data.
Source: VOSviewer.

Table 9.3. Themes Based on Text-Data Analysis.

S. No.	Cluster Theme	Key Items
1	Sustainability	Agricultural practice, agricultural production, technology, environment, ML, profitability, sustainable agriculture, sensing, water
2	Research and knowledge building	Barrier, greenhouse gas emission, improvement, attitude, risk, opportunity, framework
3	Technology	AI, big data, cloud, IoT, robotic
4	Measures	Agricultural sustainability, climate change, crop production, fertilization, life cycle assessment
5	Solutions	Energy, pesticide, product, remote sensing

Source: Items from VOSviewer and nomenclature for themes by authors.

3. Discussion and Implications

In this section, we outline the key insights from the BMO framework as shown in Fig. 9.4. We note that, with the increasing demand for food and decreasing interest of farmers in agriculture occupation, there is a need for a massive increase in farm productivity and farmers profitability fulfilling the ecological and social requirements (Harisudin et al., 2023; Islam et al., 2021; Ouafiq et al., 2022; Ragazou et al., 2022; Yadav et al., 2022).

Therefore, future scholars should focus on how latest technologies such as ML can transform the existing framing systems into data-driven smart farming systems? More studies on how ML algorithms can contribute towards facilitating the optimal use and utilization of resources throughout the smart technology are required to be conducted (Orjuela-Garzon et al., 2021; Osrof et al., 2023; Quintero et al., 2022; Tama et al., 2023).

There are a lot of expectations from the local government by the research members. Considering the high cost of investment in digital technologies and developing ML capabilities, the policymakers are expected to subsidize investments in digital technologies and make it more affordable so that it can be used widely (Akbar et al., 2024; Balkrishna et al., 2023; Ennouri & Kallel, 2019; Javaid et al., 2022). Device interoperability is a significant technical barrier that would be needed to overcome for ensuring the penetration of ML techniques and government aid.

It is, therefore, evident that the practitioners will have to plan new investments in emerging technologies to solve the social and environmental problems, simultaneously increasing efficiency. Apart from the sustainability benefits, the ML applications also contribute to improving supply chain visibility, transparency and product traceability. However, this will require the practitioners to explore the possibilities of integrating different sources of ML data with blockchain technology and other technologies.

The ML algorithms increase the likelihood of early detection of food safety-related issues. ML algorithms based on simulation technologies are found to support early mitigation of risks. It is therefore implied that the practitioners should invest in sensor technologies for monitoring the real-time processing activities, ensuring machine interoperability and real-time decision-making (Eastwood et al., 2019; Jakku et al., 2019; Knierim et al., 2019).

The machine vision applications combined with ML algorithms can be beneficial in predicting plant stress conditions, which will help in taking proactive decisions on the target plants without harming other plants in the vicinity. This will help in the optimal use of resources such as fertilizers, water, pesticides and weedicides (Atanasov, 2023; Pero et al., 2024).

One significant non-technical barrier in this field is the lack of policy regulations on the use of data, the involvement of different stakeholders and information sharing practices. Further, there is a shortage of human resources with analytical skills that may affect the adoption of ML applications (Akbar et al., 2024; Balkrishna et al., 2023; Ennouri & Kallel, 2019; Javaid et al., 2022). In the emerging markets, there exists a considerable gap between the available resources and education. Many farmers do not have access to the internet, mobile phones

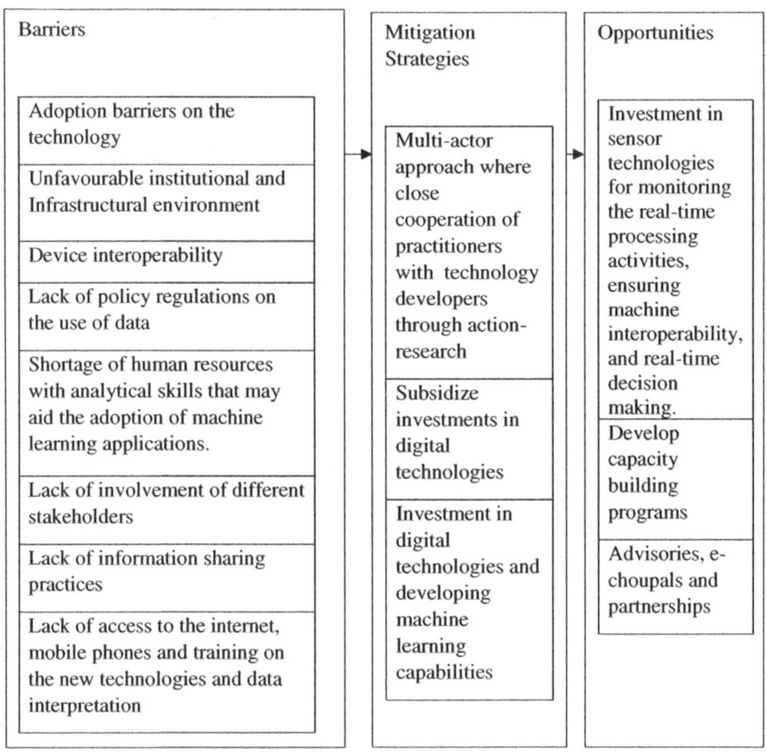

Fig. 9.4. BMO Framework in the Context of Sustainability and Smart Farming in India. *Source*: Authors' own.

and training on the new technologies and data interpretation. Advisories are required to be developed to assist the farmers in understanding the information and recommend suitable mechanisms to improve farm productivity (Harisudin et al., 2023; Islam et al., 2021). It is therefore implied that the policymakers should plan to connect the farmers in the rural areas and work with governments and technology companies for pulling costs of data collection equipment and software. There is also the need to provide extensive and advanced education, which revolves around utilizing these farming measures globally. Therefore, it is required to develop capacity-building programmes for enhancing the skill set of the farmers and other stakeholders (Orjuela-Garzon et al., 2021; Osrof et al., 2023).

4. Conclusion and Limitation

We can conclude that with technology, farming has become data driven, enabling timely, cost-effective and ecologically sustainable farm management. Since India lack the infrastructure for precision farming; hence, a methodological paradigm is needed to connect new technology for cost-effectively, considering the existing limits and constraints. Crop yields are lowered by natural disasters, peasants, careless seed planting, erratic watering, and locust infestations. Farmers face significant challenges due to soil erosion, biodiversity loss and increased food demand. Farmers often give up and accept less than they deserve when faced with difficulties of this nature. It is time for the Indian government to aid farmers in the adoption of smart farming practices and technology to keep constant tabs on their crops and turn forth delicious food. In this work, the authors conducted a comprehensive literature review utilizing bibliometric tools. We identified five distinct themes where smart farming and farming are currently progressing. Trends in these areas can help and guide the efforts of present and future researchers.

As research in this area is emerging actively, we have attempted a comprehensive review of the field, yet we face a sample-based limitation, as this research only comprises articles published only in journals and in English language; thus, future studies can include languages other than English books, conference papers, and editorials to generate broader understanding of the field.

Nevertheless, the findings of this study have the potential to serve as a guide for researchers who wish to explore this burgeoning field of smart farming and sustainability in India.

References

Akbar, J. U. M., Kamarulzaman, S. F., Muzahid, A. J. M., Rahman, Md. A., & Uddin, M. (2024). A comprehensive review on deep learning assisted computer vision techniques for smart greenhouse agriculture. *IEEE Access*, *12*, 4485–4522. https://doi.org/10.1109/ACCESS.2024.3349418

Atanasov, S. (2023). State-of-the-art technologies for remote sensing of crops water status and nutrients in agriculture: A review. *Scientific Horizons*, *26*(9), 167–177. https://doi.org/10.48077/scihor9.2023.167

Balkrishna, A., Pathak, R., Kumar, S., Arya, V., & Singh, S. K. (2023). A comprehensive analysis of the advances in Indian digital agricultural architecture. *Smart Agricultural Technology*, *5*, 100318. https://doi.org/10.1016/j.atech.2023.100318

Bathla, A., Chawla, G., & Gupta, A. (2024). Benchmarking design-thinking as a tool for education: A systematic review and future research agenda. *Benchmarking: An International Journal*. https://doi.org/10.1108/BIJ-09-2023-0603

Bathla, A., Chawla, G., Hofaidhllaoui, M., & Dabic, M. (2024). Exploring the dynamics of design thinking in management education and training: A critical review, taxonomic analysis and practical implications. *European Journal of Innovation Management*, *27*(9), 337–359. https://doi.org/10.1108/EJIM-12-2023-1108

Eastwood, C., Ayre, M., Nettle, R., & Dela Rue, B. (2019). Making sense in the cloud: Farm advisory services in a smart farming future. *NJAS: Wageningen Journal of Life Sciences*, *90–91*(1), 1–10. https://doi.org/10.1016/j.njas.2019.04.004

Ennouri, K., & Kallel, A. (2019). Remote sensing: An advanced technique for crop condition assessment. *Mathematical Problems in Engineering*, *2019*, 1–8. https://doi.org/10.1155/2019/9404565

Harisudin, M., Kusnandar, Riptanti, E. W., Setyowati, N., & Khomah, I. (2023). Determinants of the internet of things adoption by millennial farmers. *AIMS Agriculture and Food*, *8*(2), 329–342. https://doi.org/10.3934/agrfood.2023018

Islam, N., Rashid, M. M., Pasandideh, F., Ray, B., Moore, S., & Kadel, R. (2021). A review of applications and communication technologies for internet of things (IoT) and unmanned aerial vehicle (UAV) based sustainable smart farming. *Sustainability*, *13*(4), 1821. https://doi.org/10.3390/su13041821

Jakku, E., Taylor, B., Fleming, A., Mason, C., Fielke, S., Sounness, C., & Thorburn, P. (2019). "If they don't tell us what they do with it, why would we trust them?" Trust, transparency and benefit-sharing in Smart Farming. *NJAS: Wageningen Journal of Life Sciences*, *90–91*(1), 1–13. https://doi.org/10.1016/j.njas.2018.11.002

Javaid, M., Haleem, A., Singh, R. P., & Suman, R. (2022). Enhancing smart farming through the applications of Agriculture 4.0 technologies. *International Journal of Intelligent Networks*, *3*, 150–164. https://doi.org/10.1016/j.ijin.2022.09.004

Knierim, A., Kernecker, M., Erdle, K., Kraus, T., Borges, F., & Wurbs, A. (2019). Smart farming technology innovations – Insights and reflections from the German Smart-AKIS hub. *NJAS: Wageningen Journal of Life Sciences*, *90–91*(1), 1–10. https://doi.org/10.1016/j.njas.2019.100314

Orjuela-Garzon, W., Quintero, S., Giraldo, D. P., Lotero, L., & Nieto-Londoño, C. (2021). A theoretical framework for analysing technology transfer processes using agent-based modelling: A case study on massive technology adoption (AMTEC) program on rice production. *Sustainability*, *13*(20), 11143. https://doi.org/10.3390/su132011143

Osrof, H. Y., Tan, C. L., Angappa, G., Yeo, S. F., & Tan, K. H. (2023). Adoption of smart farming technologies in field operations: A systematic review and future research agenda. *Technology in Society*, *75*, 102400. https://doi.org/10.1016/j.techsoc.2023.102400

Ouafiq, E. M., Saadane, R., & Chehri, A. (2022). Data management and integration of low power consumption embedded devices IoT for transforming smart agriculture into actionable knowledge. *Agriculture*, *12*(3), 329. https://doi.org/10.3390/agriculture12030329

Pero, C., Bakshi, S., Nappi, M., & Tortora, G. (2024). IoT-driven machine learning for precision viticulture optimization. *IEEE Journal of Selected Topics in Applied Earth Observations and Remote Sensing*, *17*, 2437–2447. https://doi.org/10.1109/JSTARS.2023.3345473

Quintero, S., Giraldo, D. P., & Garzon, W. O. (2022). Analysis of the specialization patterns of an agricultural innovation system: A case study on the banana production chain (Colombia). *Sustainability*, *14*(14), 8550. https://doi.org/10.3390/su14148550

Ragazou, K., Garefalakis, A., Zafeiriou, E., & Passas, I. (2022). Agriculture 5.0: A new strategic management mode for a cut cost and an energy efficient agriculture sector. *Energies*, *15*(9), 3113. https://doi.org/10.3390/en15093113

Tama, R. A. Z., Hoque, M. M., Liu, Y., Alam, M. J., & Yu, M. (2023). An application of partial least squares structural equation modeling (PLS-SEM) to examining farmers' behavioral attitude and intention towards conservation agriculture in Bangladesh. *Agriculture*, *13*(2), 503. https://doi.org/10.3390/agriculture13020503

Yadav, S., Kaushik, A., Sharma, M., & Sharma, S. (2022). Disruptive technologies in smart farming: An expanded view with sentiment analysis. *AgriEngineering*, *4*(2), 424–460. https://doi.org/10.3390/agriengineering4020029

Chapter 10

Analyzing Spatial and Temporal Variability of Climatic Variables in Uttar Pradesh: Implications for Indian Agriculture

Srishty Kasana[a], *Yamini Gupt*[b], *Amarnath Tripathi*[c] *and Parv Kasana*[d]

[a]*Daulat Ram College, University of Delhi (North Campus), Delhi, India*
[b]*Department of Finance and Business Economics, University of Delhi (South Campus), Delhi, India*
[c]*Jaipuria Institute of Management, Noida, India*
[d]*Department of Geology, University of Delhi (North Campus), Delhi, India*

Abstract

This study examines the spatial and temporal variability of climatic variables, particularly surface temperature (both minimum and maximum) and rainfall, in Uttar Pradesh over a period of 118 years (1900–2018). Using climate data and applying regression analysis and the Mann–Kendall (MK) test, the findings reveal significant increases in both minimum and maximum temperatures across most districts. The analysis shows that the minimum temperature in 2018 ranged from 20.1°C to 24.5°C, compared to 17.9°C–19.5°C in 1900, with nearly all districts experiencing a rise. Similarly, the maximum temperature increased from 31.1°C–31.5°C in 1900 to 31.2°C–32.1°C in 2018. Rainfall variability also exhibited changes, with certain regions experiencing shifts in patterns, though no significant upward trend was observed. The results highlight the potential impact of these climatic changes on agriculture, particularly rainfed crops like wheat, and underscore the need for effective adaptation strategies in the face of climate change.

Keywords: Spatial and temporal variability; meteorological variables; climate change; Mann–Kendall test; Uttar Pradesh

Sustainable and Smart Farming Eco-systems:
Changing Landscapes in the Age of Technology, 129–145

doi:10.1108/978-1-83608-092-320261010

1. Introduction

Climate change is a global phenomenon that severely impacts both lives and livelihoods worldwide. It has been recognized as one of the greatest threats to development, necessitating immediate policy action. The literature suggests two primary responses to tackle the problem of climate change: mitigation and adaptation. Mitigation has received more attention since the beginning; however, very encouraging outcomes have yet to be realized, primarily due to its interdependent nature. The success of a country's initiatives to mitigate greenhouse gas emissions largely depends on the actions of other countries. In contrast, adaptation is independent in nature and is crucial because it helps reduce the adverse impacts of greenhouse gases already present in the atmosphere. However, the effectiveness of adaptation responses mainly relies on local actions. Therefore, understanding climate change at the local level is critical.

Against this backdrop, the present study aims to explore month-wise trends in temperature and rainfall variability across districts in Uttar Pradesh, providing critical insights for targeted policy interventions. Uttar Pradesh's pivotal role in India's food grain production means that any climate-related threat to its agricultural output not only endangers the state but also poses a significant risk to the nation's overall food security. This highlights the necessity of conducting a district-level trend analysis to identify climatic variables showing significant changes, enabling policymakers to prioritize specific districts in designing climate resilience strategies. The study focuses on district-level trends, rather than state-level aggregation, to avoid masking localized variations in climatic data. Moreover, it examines a substantial time span of 118 years, providing robust insights into long-term trends and their implications.

This research primarily employs the MK test to detect trends in climatic variables. The MK test has been widely applied in numerous studies, including those by Gadedjisso et al. (2021), Meena (2020), Jayanta et al. (2020), Kaur et al. (2017), Kumar et al. (2010), Hussain et al. (2015), Chinchorkar et al. (2015), Palanichamy et al. (2020), Padhiary et al. (2018), Salehi et al. (2020), Ataei et al. (2013), Mahajan and Dodamani (2015), Sinha et al. (2019), Pingale et al. (2016), Sulaiman et al. (2015), Kişi et al. (2018), Ganguly et al. (2015), Yilmaz (2019), Sasane (2016), Asfaw (2016), and Sridhar and Raviraj (2017).

2. Data and Methods

Data on rainfall and temperature were retrieved primarily from three sources: India Water Portal, World Weather Online, and the India Meteorological Department. To identify trends, data on rainfall and temperature spanning 118 years (1900–2018) were analyzed. Huth and Pokorná (2004) argue that assessing long-term variations in climate variables is an essential step in detecting climate change. Statistical methods commonly used in climatology include estimating moments of statistical distributions, analyzing trends, and performing correlation analysis. However, before employing these parametric tests, certain assumptions must be met, such as the type of distribution (most frequently normal), serial independence (i.e., zero autocorrelation), and stationarity. In many cases, climatological data violate these assumptions. An alternative solution is offered by non-parametric tests (distribution-free methods). Non-parametric tests are better suited to handling outliers compared to parametric tests, though they are less efficient. This inefficiency arises because they produce greater uncertainty in statistical estimates. Yue et al. (2002) highlight that the MK test is often preferred over parametric tests for analyzing hydro-meteorological time series due to its suitability for non-normally distributed and censored data – issues frequently encountered in such data sets. Therefore, the MK test (Kendall, 1975; Mann, 1945) was applied to detect statistically significant trends in climatic variables for Uttar Pradesh. The null hypothesis (H_0) posits that there is no trend in precipitation over time, while the alternative hypothesis (H_1) asserts the presence of a trend (either increasing or decreasing). The test uses the following equations to calculate the MK statistic (S), its variance ($V(S)$), and the standardized test statistic (Z):

$$S = \sum_{k=1}^{n-1} \sum_{j=k+1}^{n} \text{sgn}(x_j - x_k) \tag{1}$$

$$\text{sgn}(x_j - x_k) = \begin{Bmatrix} +1 & \text{if } x_j - x_k > 0 \\ 0 & \text{if } x_j - x_k = 0 \\ -1 & \text{if } x_j - x_k < 0 \end{Bmatrix} \tag{2}$$

where x_j and x_k are values in periods j and k, respectively, and given $j > k$. n is the number of observed data series. If the value of S is greater than 0, it indicates the increasing trend, and if the value is less than 0, it indicates a decreasing trend. In order to establish the quantifiable statistical significance of the trend mentioned above, Kendall introduced a normal-approximation test for those data sets which have more than 10 values, and there are not many tied values within the data set. The normalized test statistics Z is as follows:

$$Z = \begin{Bmatrix} \dfrac{S-1}{\sqrt{\text{VAR}(S)}} & \text{if } S > 0 \\ 0 & \text{if } S = 0 \\ \dfrac{S+1}{\sqrt{\text{VAR}(S)}} & \text{if } S < 0 \end{Bmatrix} \tag{3}$$

Variance for statistics S is determined as:

$$\text{VAR}(S) = \frac{1}{18}\left[n(n-1)(2n+5) - \int_{p=1}^{q} t_p(t_p - 1)(2t_p + 5) \right] \tag{4}$$

where q is the number of tied groups and t_p is the number of data values in the pth group. The $Z_{1\text{-}\alpha/2}$ is the critical value of Z from the standard normal table; for 90% level of confidence, the value of $Z_{1\text{-}\alpha/2}$ is 1.64. If $|Z|>Z_{1\text{-}\alpha/2}$, we fail to accept the null hypothesis, and there is an existence of a significant trend in the time series.

Like mentioned above, if $Z > 0$, there exists an upward trend, and if the value of Z is less than 0, it indicates a downward trend in the variable.

Non-parametric tests help identify trends in temperature and rainfall series. However, to capture the magnitude of changes in these series, parametric tests were also employed. Hirsch et al. (as cited in Tabari et al., 2011) note that simple linear regression (SLR) is a widely used parametric technique for analyzing monotonic trends in time series. SLR investigates the relationship between a dependent variable and time, often estimating the slope of climatic or hydrological variables over time. A positive slope (b) indicates an upward trend, while a negative slope reflects a downward trend. SLR also provides a measure of significance through hypothesis testing on the slope, as well as the magnitude of the rate of change. The regression equation is given as:

$$Y_t = a + bt + e_t \cdots \tag{5}$$

where Y_t = climatic variable (e.g., minimum temperature, maximum temperature, or rainfall); b = slope of the regression equation; t = time (in years); e = error term; and a = regression constant.

3. Results and Discussion

Focusing on the spatial and temporal variability of meteorological variables in the context of climate variations is crucial. When agriculture in specific regions depends heavily on rainfall, these variables play a vital role in understanding climate-induced changes. Given the prevalence of rainfed agriculture in many developing nations, researchers over the past few decades have increasingly emphasized the importance of understanding the spatial and temporal variability of rainfall. These studies recommend feasible adaptation strategies tailored to specific regions (Rushi Kumar et al., 2019). Mahmood et al. (2019) highlight the critical role of accurate forecasts of climatic variables in assisting policymakers, planners, and other stakeholders in water resource management, mitigation, and adaptation efforts to combat the adverse effects of climate change. Accurate predictions can facilitate efficient management of water demand and supply, mitigate droughts and floods, maintain water reservoir levels, develop precise disaster response plans, reduce uncertainties related to future water availability, and optimize water resource allocation.

Observations reveal that all seven agro-climatic zones of Uttar Pradesh are experiencing an upward trend in both maximum and minimum temperatures. Table 10.1 provides a comprehensive list of districts exhibiting upward trends in minimum temperature, maximum temperature, and rainfall variability. However, certain districts show a downward trend (albeit statistically insignificant) in maximum temperatures during specific months. For instance, districts such as Azamgarh, Auraiyya, Allahabad, Baghpat, Bahraich, Ballia, Balrampur, Barabanki, Basti, Bijnor, Budaun, Bulandsheher (which also showed a downward trend in October), Deoria, Faizabad, Farrukhabad, Ghazipur, Gorakhpur, Jaunpur, Kheri, Mainpuri, Mathura, Mau, Saharanpur, Shahjahanpur, Pratapgarh,

Table 10.1 Districts Showing Upward Trend in Rainfall and Temperature Series.

Month	Minimum Temperature	Maximum Temperature	Rainfall Variability
January	Banda, Bahraich, Barabanki, Basti, Bijnor, Farrukhabad, Kheri Moradabad, Saharanpur, Sitapur, Agra, Aligarh, Allahabad, Auraiyya, Azamgarh, Baghpat, Bahraich, Balrampur, Budaun, Bulandsheher, Deoria, Etah, Etawah, Faizabad, Fatehpur, Firozabad, Ghaziabad, Ghazipur, Gonda, Gorakhpur, Hamirpur, Hardoi, Jhansi, Lalitpur, Lucknow, Mainpuri, Mathura, Mau, Meerut, Mirzapur, Rampur, Jalauan, Shahjahanpur, Varanasi, Hathras, Kannauj , Mahoba, Pilibhit, Sultanpur, Unnao, Jaunpur, Pratapgarh, Maharaj Ganj	Bahraich, Barabanki, Basti, Bijnor, Farrukhabad, Kheri Moradabad, Pilibhit Saharanpur, Sitapur	Basti
February	Ballia, Banda, Bahraich, Barabanki, Basti, Bijnor, Farrukhabad, Kheri, Moradabad, Saharanpur, Sitapur, Agra, Aligarh, Allahabad, Auraiyya, Azamgarh, Baghpat, Bahraich, Balrampur, Budaun, Bulandsheher, Deoria, Etah, Etawah, Faizabad, Fatehpur, Firozabad, Ghaziabad, Ghazipur, Gonda, Gorakhpur, Hamirpur, Hardoi, Jhansi Lalitpur, Lucknow, Mainpuri, Mathura, Mau, Meerut, Mirzapur, Rampur, Jalauan, Shahjahanpur, Varanasi, Hathras, Kannauj , Mahoba, Pilibhit, Sultanpur, Unnao, Jaunpur, Pratapgarh, Maharaj Ganj	Bahraich, Barabanki, Basti, Bijnor, Farrukhabad, Kheri, Moradabad, Saharanpur, Sitapur, Agra, Aligarh, Allahabad, Auraiyya, Azamgarh, Baghpat, Jalauan, Bahraich, Balrampur, Budaun, Bulandsheher, Deoria, Etah, Etawah, Faizabad, Fatehpur, Firozabad, Ghaziabad, Ghazipur, Gonda, Gorakhpur, Hamirpur, Hardoi, Jhansi, Lalitpur, Lucknow, Mainpuri, Mathura, Mau, Meerut, Mirzapur, Rampur, Shahjahanpur, Varanasi, Hathras, Kannauj , Mahoba, Pilibhit, Sultanpur, Unnao, Jaunpur, Pratapgarh, Maharaj Ganj, Jalaun	Ballia, Basti, Faizabad

(Continued)

Table 10.1 (*Continued*)

Month	Minimum Temperature	Maximum Temperature	Rainfall Variability
March	Banda, Bahraich, Barabanki, Basti, Bijnor, Farrukhabad, Kheri, Moradabad, Sitapur, Agra, Aligarh, Allahabad, Auraiyya, Azamgarh, Bahraich, Balrampur, Budaun, Bulandsheher, Deoria, Etah, Etawah, Faizabad, Fatehpur, Firozabad, Ghaziabad, Ghazipur, Gonda, Gorakhpur, Hamirpur, Hardoi, Jhansi, Lalitpur, Lucknow, Mainpuri, Mathura, Mau, Mirzapur, Rampur, Jalauan, Shahjahanpur, Varanasi, Hathras, Kannauj , Mahoba, Pilibhit, Sultanpur, Unnao, Jaunpur, Pratapgarh, Maharaj Ganj	Ballia, Banda, Bahraich, Barabanki, Basti, Bijnor, Farrukhabad, Kheri, Moradabad, Saharanpur, Sitapur, Agra, Aligarh, Allahabad, Auraiyya, Azamgarh, Baghpat, Bahraich, Balrampur, Budaun, Bulandsheher, Deoria, Etah, Etawah, Faizabad, Fatehpur, Firozabad, Ghaziabad, Ghazipur, Gonda, Gorakhpur, Hamirpur, Hardoi, Jhansi, Lalitpur, Lucknow, Mainpuri, Mathura, Mau, Meerut, Mirzapur, Rampur, Jalauan, Shahjahanpur, Varanasi, Hathras, Kannauj , Mahoba, Pilibhit, Sultanpur, Unnao, Jaunpur, Pratapgarh, Maharaj Ganj	Basti, Budaun, Ghaziabad
April	Banda, Bahraich, Barabanki, Basti, Bijnor, Farrukhabad, Kheri, Moradabad, Saharanpur, Sitapur, Agra, Aligarh, Allahabad, Auraiyya, Azamgarh, Baghpat, Bahraich, Balrampur, Budaun, Bulandsheher, Deoria, Etah, Etawah, Faizabad, Fatehpur, Firozabad, Ghaziabad, Ghazipur, Gonda, Gorakhpur, Hamirpur, Hardoi, Jhansi, Lalitpur, Lucknow, Mainpuri, Mathura, Mau, Meerut, Mirzapur, Rampur, Jalauan, Shahjahanpur, Varanasi, Hathras, Kannauj , Mahoba, Pilibhit, Sultanpur, Unnao, Jaunpur, Pratapgarh, Maharaj Ganj	Ballia, Banda, Bahraich, Barabanki, Basti, Bijnor, Farrukhabad, Kheri, Moradabad, Saharanpur, Sitapur, Agra, Aligarh, Allahabad, Auraiyya, Azamgarh, Baghpat, Bahraich, Balrampur, Budaun, Bulandsheher, Deoria, Etah, Etawah, Faizabad, Fatehpur, Firozabad, Ghaziabad, Ghazipur, Gonda, Gorakhpur, Hamirpur, Hardoi, Jhansi, Lalitpur, Lucknow, Mainpuri, Mathura, Mau, Meerut, Mirzapur, Rampur, Shahjahanpur, Varanasi, Hathras, Kannauj , Mahoba, Pilibhit, Sultanpur, Unnao, Jaunpur, Pratapgarh, Jalauan	

May	Ballia, Banda, Bahraich, Barabanki, Basti, Bijnor, Farrukhabad, Kheri, Moradabad, Saharanpur, Sitapur, Agra, Aligarh, Allahabad, Auraiyya, Azamgarh, Baghpat, Bahraich, Balrampur, Budaun, Bulandsheher, Deoria, Etah, Etawah, Faizabad, Fatehpur, Firozabad, Ghaziabad, Ghazipur, Gonda, Gorakhpur, Hamirpur, Hardoi, Jhansi, Lalitpur, Lucknow, Mainpuri, Mathura, Mau, Meerut, Mirzapur, Rampur, Jalauan, Shahjahanpur, Varanasi, Hathras, Kannauj , Mahoba, Pilibhit, Sultanpur, Unnao, Jaunpur, Pratapgarh, Maharaj Ganj	Bahraich, Barabanki, Basti, Bijnor, Farrukhabad, Kheri, Moradabad, Saharanpur, Sitapur, Ballia, Agra, Aligarh, Allahabad, Auraiyya, Azamgarh, Baghpat, Bahraich, Balrampur, Banda, Budaun, Bulandsheher, Deoria, Etah, Etawah, Faizabad, Fatehpur, Firozabad, Ghaziabad, Ghazipur, Gonda, Gorakhpur, Hamirpur, Hardoi, Jhansi, Lalitpur, Lucknow, Mainpuri, Mathura, Mau, Meerut, Mirzapur, Rampur, Shahjahanpur, Varanasi, Hathras, Kannauj , Mahoba, Jalauan	Agra, Aligarh, Allahabad, Baghpat, Bahraich, Ballia, Banda, Bijnor, Deoria, Etah, Faizabad, Fatehpur, Firozabad, Ghazipur, Gorakhpur, Hamirpur, HardoiKheri, Lucknow, Meerut, Pratapgarh, Sitapur, Varanasi
June	Ballia, Bahraich, Allahabad, Basti, Bijnor, Kheri, Sitapur, Azamgarh, Bahraich, Balrampur, Deoria, Faizabad, Ghazipur, Gonda, Gorakhpur, Hardoi, Lalitpur, Lucknow, Mau, Mirzapur, Varanasi, Mahoba, Sultanpur, Jaunpur, Pratapgarh, Maharaj Ganj	Azamgarh, Ballia, Balrampur, Deoria, Ghazipur, Gonda, Gorakhpur, Maharaj Ganj, Mau	Aligarh, Baghpat,, Ballia, Balrampur, Deoria, Fatehpur, Pilibhit, Rampur

(*Continued*)

Table 10.1 (*Continued*)

Month	Minimum Temperature	Maximum Temperature	Rainfall Variability
July	Ballia, Banda, Bahraich, Basti, Farrukhabad, Sitapur, Agra, Allahabad, Auraiyya, Azamgarh, Bahraich, Balrampur, Deoria, Etawah, Faizabad, Fatehpur, Ghaziabad, Ghazipur, Gonda, Gorakhpur, Hamirpur, Hardoi, Jhansi, Lalitpur, Lucknow, Mathura, Mau, Meerut, Mirzapur, Jalauan, Shahjahanpur, Varanasi, Mahoba, Jaunpur, Pratapgarh, Maharaj Ganj	Allahabad, Ballia, Balrampur, Banda, Ghazipur, Hamirpur, Jalauan, Lalitpur, Mau, Deoria, Mirzapur, Varanasi, Mahoba.	Agra, Allahabad, Auraiyya, Azamgarh, Bahraich, Ballia, Balrampur, Banda, Barabanki, Basti, Deoria, Faizabad
August	Banda, Bahraich, Barabanki, Basti, Agra, Aligarh, Allahabad, Auraiyya, Azamgarh, Bahraich, Balrampur, Bulandsheher, Deoria, Etawah, Faizabad, Fatehpur, Firozabad, Ghaziabad, Ghazipur, Gonda, Gorakhpur, Hamirpur, Jhansi, Lalitpur, Mathura, Mau, Mirzapur, Jalauan, Varanasi, Hathras, Mahoba, Maharaj Ganj	Allahabad, Banda, Ghaziabad, Lalitpur, Mathura, Varanasi, Mahoba, Jhansi.	Allahabad, Auraiyya, Azamgarh, Ballia, Balrampur, Banda, Barabanki, Basti, Bijnor, Bulandsheher, Deoria, Etah, Etawah, Faizabad
September	Banda, Jhansi, lalitpur, Maharaj Ganj, Mathura, Mau,	Nil	Balrampur, Mau, Fatehpur, Jaunpur, Jhansi, Lalitpur, Varanasi

October	Banda, Barabanki, Basti, Farrukhabad, Sitapur, Agra, Aligarh, Allahabad, Auraiyya, Azamgarh, Bahraich, Budaun, Bulandsheher, Deoria, Etah, Etawah, Faizabad, Fatehpur, Firozabad, Ghaziabad, Ghazipur, Gorakhpur, Hamirpur, Hardoi, Jhansi, Lalitpur, Lucknow, Mainpuri, Mathura, Mirzapur, Jalauan, Shahjahanpur, Varanasi, Hathras, Kannauj, Mahoba, Sultanpur, Unnao, Jaunpur, Pratapgarh, Maharaj Ganj	Mahoba, Lalitpur, Mirzapur, Jhansi, Budaun, Bulandsheher, Farrukhabad, Banda	Lucknow
November	Ballia, Banda, Bahraich, Barabanki, Basti, Bijnor, Farrukhabad, Kheri, Moradabad, Saharanpur, Sitapur, Agra, Aligarh, Allahabad, Auraiyya, Azamgarh, Baghpat, Bahraich, Balrampur, Budaun, Bulandsheher, Deoria, Etah, Etawah, Faizabad, Fatehpur, Firozabad, Ghaziabad, Ghazipur, Gonda, Gorakhpur, Hamirpur, Hardoi, Jhansi, Lalitpur, Lucknow, Mainpuri, Mathura, Mau, Meerut, Mirzapur, Rampur, Jalauan, Shahjahanpur, Varanasi, Hathras, Kannauj , Mahoba, Pilibhit, Sultanpur, Unnao, Jaunpur, Pratapgarh, Maharaj Ganj	Ballia, Banda, Bahraich, Barabanki, Farrukhabad, Moradabad, Sitapur, Aligarh, Allahabad, Auraiyya, Azamgarh, Baghpat, Deoria, Etawah, Faizabad, Fatehpur, Firozabad, Ghaziabad, Ghazipur, Gonda, Gorakhpur, Hamirpur, Hardoi, Jhansi, Lalitpur, Mainpuri, Mathura, Mau, Meerut, Mirzapur, Jalauan, Varanasi, Kannauj , Mahoba, Sultanpur, Unnao, Jaunpur, Pratapgarh	Balrampur, Ghaziabad, Gonda, Sitapur

(*Continued*)

Table 10.1 (*Continued*)

Month	Minimum Temperature	Maximum Temperature	Rainfall Variability
December	Ballia, Banda, Bahraich, Barabanki, Basti, Bijnor, Farrukhabad, Kheri, Moradabad, Saharanpur, Sitapur, Agra, Aligarh, Allahabad, Auraiyya, Azamgarh, Baghpat, Bahraich, Balrampur, Budaun, Bulandsheher, Deoria, Etah, Etawah, Faizabad, Fatehpur, Firozabad, Ghaziabad, Ghazipur, Gonda, Gorakhpur, Hamirpur, Hardoi, Jhansi, Lalitpur, Lucknow, Mainpuri, Mathura, Mau, Meerut, Mirzapur, Rampur, Jalauan, Shahjahanpur, Varanasi, Hathras, Kannauj , Mahoba, Pilibhit, Sultanpur, Unnao, Jaunpur, Pratapgarh, Maharaj Ganj	Ballia, Banda, Bahraich, Barabanki, Basti, Bijnor, Farrukhabad, Kheri, Moradabad, Saharanpur, Sitapur, Agra, Aligarh, Allahabad, Auraiyya, Azamgarh, Baghpat, Bahraich, Balrampur, Budaun, Bulandsheher, Deoria, Etah, Etawah, Faizabad, Fatehpur, Firozabad, Ghaziabad, Ghazipur, Gonda, Gorakhpur, Hamirpur, Hardoi, Jhansi, Lalitpur, Lucknow, Mainpuri, Mathura, Mau, Meerut, Mirzapur, Rampur, Jalauan, Shahjahanpur, Varanasi, Hathras, Kannauj , Mahoba, Pilibhit, Sultanpur, Unnao, Jaunpur, Pratapgarh, Maharaj Ganj	

Source: Authors' own calculation based on climate data collected from different sources and analyzed using MK test and regression analysis.

Rampur, Sultanpur, Unnao, and Hathras exhibited a downward trend in maximum temperatures between August and September. Agra and Aligarh showed a downward trend from June to October.

Similarly, districts such as Aligarh, Baghpat, Bahraich, Ballia, Balrampur, Bijnor, Budaun, Bulandsheher, Etah, Faizabad, Farrukhabad, Firozabad, Ghaziabad, Gonda, Hardoi, Kheri, Lucknow, Mainpuri, Meerut, Moradabad, Pilibhit, Rampur, Shahjahanpur, Unnao, Hathras, and Kannauj exhibited a downward trend (though statistically insignificant) in minimum temperatures during September. Barabanki experienced a downward trend from June to September, while Saharanpur showed a similar trend during August and September. In terms of rainfall variability, nearly all districts demonstrated a downward trend throughout the year, from January to December.

An attempt has also been made to identify districts that experienced increases in minimum average temperature, maximum temperature, and rainfall variability by comparing data from 1900 with 2018. The results are presented in Figs. 10.1–10.6.

3.1. Minimum Average Temperature

Figs. 10.1 and 10.2 illustrate the changes in minimum average temperature across districts in Uttar Pradesh. In 1900, the minimum average temperature ranged

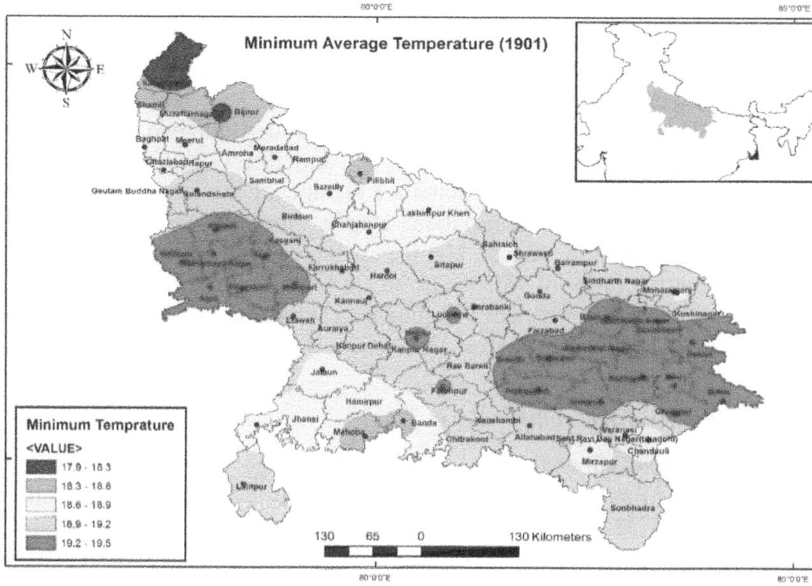

Fig. 10.1. Minimum Average Temperature 1901. *Source*: Authors' own calculation based on climate data collected from different sources and analyzed using MK test and regression analysis.*Note*: We have created these six graphs (graphs 3.2–3.7) using the software ArcGIS 10.3.1.

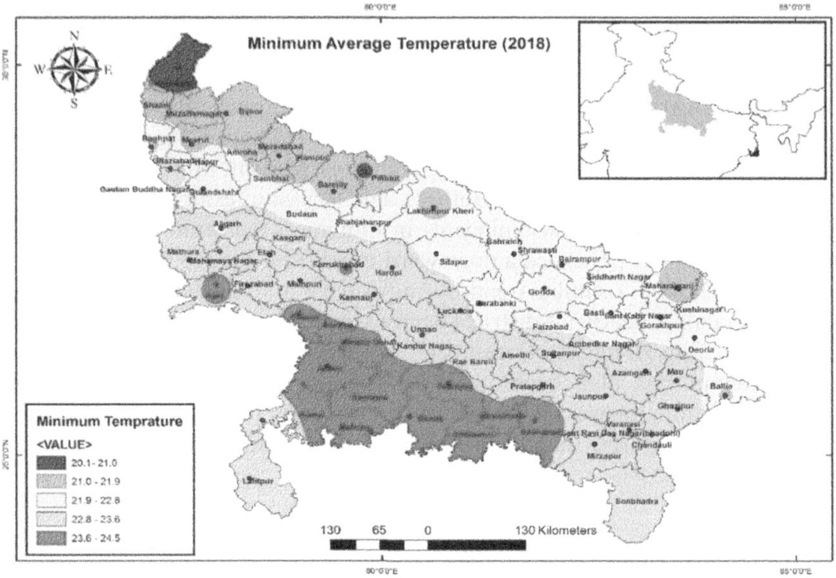

Fig. 10.2. Minimum Average Temperature 2018. *Source*: Authors' own calculation based on climate data collected from different sources and analyzed using MK test and regression analysis.

from 17.9°C to 19.5°C. By 2018, this range had increased to 20.1°C–24.5°C, with every district experiencing a rise. For instance, districts such as Etawah, Auraiyya, Jalaun, Hamirpur, Banda, Mahoba, Jhansi, and Allahabad, which had a minimum average temperature between 18.6°C and 19.2°C in 1900, recorded a range of 23.6°C–24.5°C in 2018, showing a significant increase. Similarly, districts like Baghpat, Meerut, Ghaziabad, Hapur, Amroha, Moradabad, Rampur, Shahjahanpur, Jalaun, Jhansi, Hamirpur, Banda, Mirzapur, Bahraich, and Maharajganj, with a range of 18.3°C–18.9°C in 1900, experienced an increase to 21.9°C–24.5°C by 2018. This clearly highlights a substantial upward trend in the minimum average temperature across the state.

3.2. Maximum Average Temperature

Figs. 10.3 and 10.4 show that the maximum average temperature in Uttar Pradesh ranged from 31.1°C to 31.5°C in 1900. By 2018, this range shifted slightly to 31.2°C–32.1°C. In 1900, districts such as Etah, Agra, Firozabad, Mainpuri, Kannauj, Etawah, Auraiyya, Unnao, and Fatehpur recorded the highest maximum average temperatures, ranging from 32.7°C to 33.0°C. By 2018, districts like Amroha, Moradabad, and Rampur experienced the highest maximum average temperatures, ranging from 34.6°C to 35.4°C. Additionally, districts such as Bulandshahr, Rampur, Pilibhit, Mathura, Aligarh, Budaun,

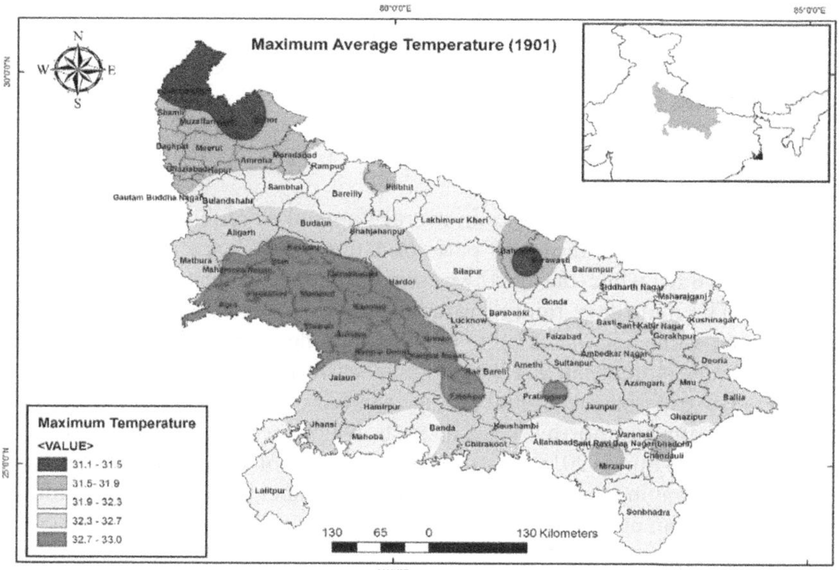

Fig. 10.3. Maximum Average Temperature 1901. *Source*: Authors' own calculation based on climate data collected from different sources and analyzed using MK test and regression analysis.

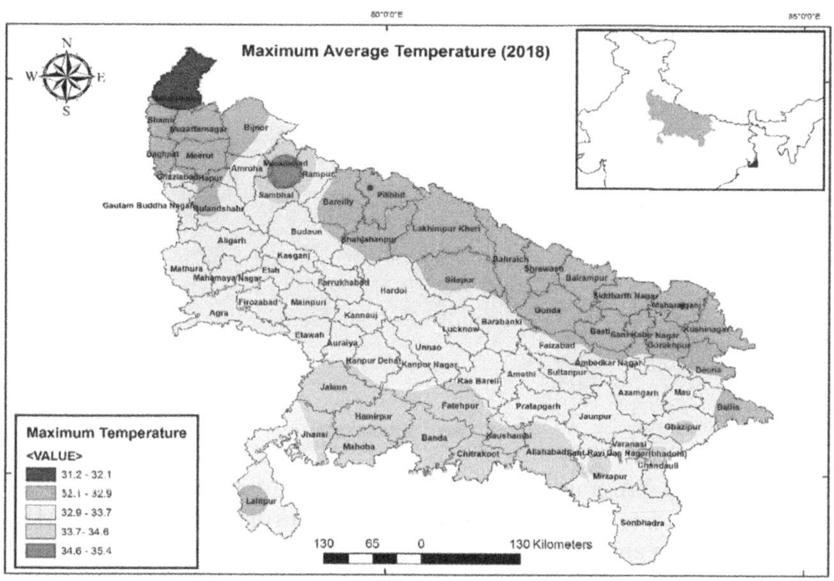

Fig. 10.4. Maximum Average Temperature 2018. *Source*: Authors' own calculation based on climate data collected from different sources and analyzed using MK test and regression analysis.

Shahjahanpur, Hardoi, Sitapur, Lucknow, Jalaun, Hamirpur, Jhansi, Mahoba, Banda, Pratapgarh, Faizabad, Sultanpur, Basti, Mau, Azamgarh, Jaunpur, Deoria, and Ghazipur experienced an increase from 31.9°C–32.7°C in 1900 to 32.9°C–34.6°C in 2018.

3.3. Rainfall Variability

Figs. 10.5 and 10.6 depict changes in average annual rainfall patterns. In 1900, the northeastern and southeastern parts of Uttar Pradesh received 77–112 mm of annual rainfall, while the northwestern and southwestern regions received 35–58 mm. The mid-range rainfall value during this period was 35–58 mm. By 2018, this mid-value had increased to 61–68 mm, reflecting significant changes in rainfall patterns. Notably, in 2018, many northeastern and southeastern districts that previously experienced annual average rainfall of 77–112 mm reported only 30–68 mm. Meanwhile, northwestern districts, which received 35–58 mm in 1900, experienced 51–88 mm in 2018. The southwestern part of Uttar Pradesh recorded annual rainfall ranging from 31 to 51 mm in 2018, indicating a noticeable shift in rainfall distribution over time.

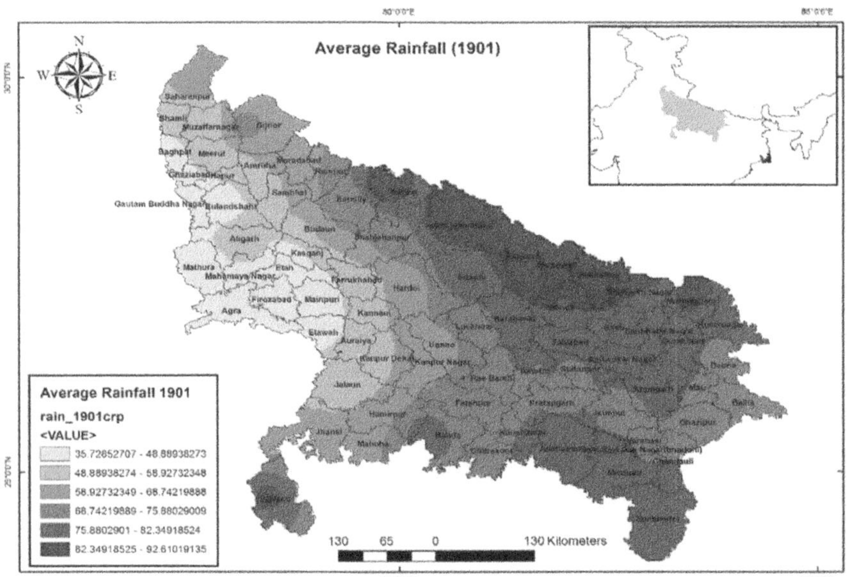

Fig. 10.5. Annual Average Rainfall 1901. *Source*: Authors' own calculation based on climate data collected from different sources and analyzed using MK test and regression analysis.

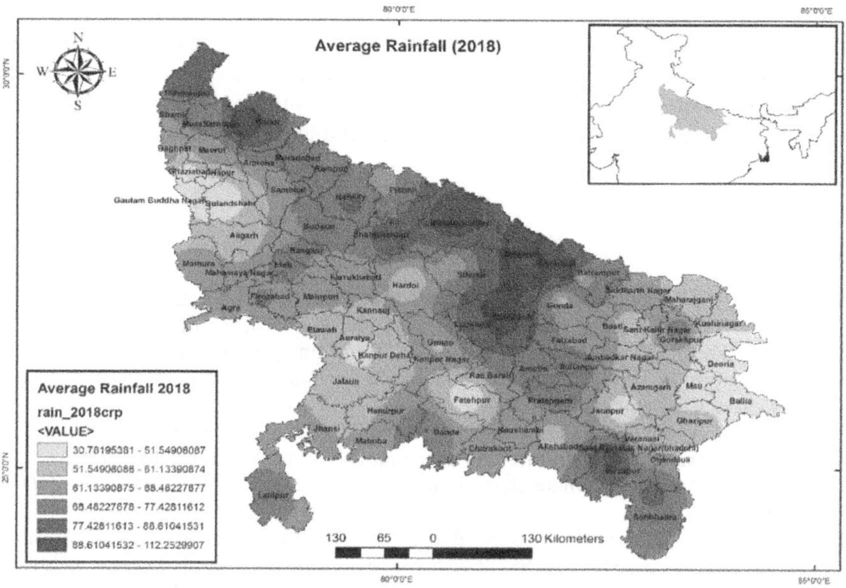

Fig. 10.6. Annual Average Rainfall 2018. *Source*: Authors' own calculation based on climate data collected from different sources and analyzed using MK test and regression analysis.

4. Conclusion

The findings of this study reveal that most districts in Uttar Pradesh have experienced significant variations in minimum and maximum temperatures over the past 118 years (1900–2018). These changes pose serious challenges to agricultural production in general and to winter crops like wheat in particular, as they are highly sensitive to temperature fluctuations. For instance, wheat requires a prolonged period of low temperatures (November–March) for optimal physiological growth, followed by moderately high temperatures during the grain-ripening stage (April). Additionally, critical stages in wheat development, such as the crown root initiation stage (20–22 days after sowing, typically in December) and the flowering and grain filling stage (March), demand sufficient water availability (Jha & Tripathi, 2017).

The results from SLR and the MK test indicate an upward trend in minimum and maximum temperatures across most districts in Uttar Pradesh, with significant increases highlighting a clear warming trend. In contrast, rainfall variability does not display a significant upward trend in most districts. However, some districts, including Ballia, Basti, Faizabad, Budaun, Ghaziabad, Agra, Aligarh, Allahabad, Baghpat, Bahraich, Banda, Bijnor, Deoria, Etah, Fatehpur, Firozabad, Ghazipur, Gorakhpur, Hamirpur, Hardoi, Kheri, Lucknow, Meerut, Pratapgarh, Sitapur, and Varanasi, exhibit an upward trend in rainfall variability.

While the direct impact of increased rainfall variability or deficit rainfall on agricultural production has been mitigated due to assured irrigation – covering more than 80% of the cropped area in Uttar Pradesh – the effects are becoming apparent in other forms. These include declining groundwater levels and rising irrigation costs, both of which pose long-term challenges to sustainable agricultural practices in the state.

References

Asfaw, A. (2016). Analysis of rainfall trend using Mann–Kendall test and the Sen's. *International Journal of Agricultural Science and Research*, *6*(2), 131–138.

Ataei, H., Hashemi, S., & Cheraghi, S. (2013). Trend analysis of the average monthly temperature in Isfahan Province. *Bulletin of Environment, Pharmacology and Life Sciences*, *3*(December), 98–102.

Chinchorkar, S. S., Sayyad, F. G., Vaidya, V. B., & Pandye, V. (2015). Trend detection in annual maximum temperature and precipitation using the Mann Kendall test – A case study to assess climate change on Anand of central Gujarat. *Mausam*, *66*(1), 1–6.

Gadedjisso, T., A., Adjegan, K. I., & Kablan, A. K. M. (2021). Rainfall and temperature trend analysis by Mann–Kendall test and significance for rainfed cereal yields in Northern Togo. *Science*, *3*(1), 17. https://doi.org/10.3390/sci3010017

Ganguly, A., Ray Chaudhuri, R., & Sharma, P. (2015). Analysis of trend of the precipitation data: A case study of Kangra District, Himachal Pradesh. *International Journal of Research – Granthaalayah*, *3*(9), 87–95. https://doi.org/10.29121/granthaalayah.v3.i9.2015.2948

Hussain, F., Nabi, G., & Boota. (2015). Rainfall trend analysis by using the Mann-Kendall test & Sen's slope estimates: A case study of District Chakwal Rain Gauge, Barani Area, Northern Punjab Province, Pakistan. *Science International-Lahore*, *27*(4), 3159–3165.

Huth, R., & Pokorná, L. (2004). Parametric versus non-parametric estimates of climatic trends. *Theoretical and Applied Climatology*, *77*(1–2), 107–112. https://doi.org/10.1007/s00704-003-0026-3

Jayanta, D., Mandal, T., Saha, P. I. U., & Bhattacharya, S. K. (2020). Variability and trends of rainfall using non-parametric approaches: A case study of semi-arid area. *Mausam*, *71*(1), 33–44.

Jha, B., & Tripathi, A. (2017). How susceptible is India's food basket to climate change. *Social Change*, *47*(1), 11–27.

Kaur, S., Diwakar, S. K., & Das, A. K. (2017). Long term rainfall trend over meteorological sub divisions and districts of India. *Mausam*, *68*(3), 439–450. https://doi.org/10.13140/RG.2.1.3908.4409

Kendall, M. G. (1975). *Rank correlation methods* (4th ed.). Charles Griffin.

Kişi, Ö., Santos, C. A. G., da Silva, R. M., & Zounemat-Kermani, M. (2018). Trend analysis of monthly streamflows using Şen's innovative trend method. *Geofizika*, *35*(1), 53–68. https://doi.org/10.15233/gfz.2018.35.3

Kumar, V., Jain, S. K., & Singh, Y. (2010). Analyse des tendances pluviométriques de long terme en Inde. *Hydrological Sciences Journal*, *55*(4), 484–496. https://doi.org/10.1080/02626667.2010.481373

Mann, H. B. (1945). Nonparametric tests against trend. *Econometrica*, *13*(3), 245–259. https://doi.org/10.2307/1907187

Mahajan, D. R., & Dodamani, B. M. (2015). Trend analysis of drought events over Upper Krishna Basin in Maharashtra. *Aquatic Procedia*, *4*(Icwrcoe), 1250–1257. https://doi.org/10.1016/j.aqpro.2015.02.163

Mahmood, R., Jia, S. & Zhu, W. (2019). Analysis of climate variability, trends, and prediction in the most active parts of the Lake Chad basin, Africa. *Scientific Reports*, *9*, 6317. https://doi.org/10.1038/s41598-019-42811-9

Meena, M. (2020). Rainfall statistical trend and variability detection using Mann-Kendall test, Sen's slope and coefficient of variance – A case study of Udaipur District (1957–2016). *Applied Ecology and Environmental Sciences*, *8*(1), 34–37. https://doi.org/10.12691/aees-8-1-5

Padhiary, J., Das, D. M., Patra, K. C., & Sahoo, B. C. (2018). Trend analysis of rainfall and temperature using the Mann Kendall test in Jaraikela catchment of Brahmani River basin. *International Journal of Agriculture Sciences*, *10*(19), 7309–7313.

Palanichamy, M., Sankaralingam, R., & Narayanasamy. (2020). Statistical studies on rainfall and time-based deviations in precipitation trends in Vaigai River Basin, TN State, India. *Indian Journal of Geo-Marine Sciences*, *49*(1), 15–23.

Pingale, S. M., Khare, D., Jat, M. K., & Adamowski, J. (2016). Trend analysis of climatic variables in an arid and semi-arid region of the Ajmer District, Rajasthan, India. *Journal of Water and Land Development*, *28*(1), 3–18. https://doi.org/10.1515/jwld-2016-0001

Rushi Kumar, M., Naveen, M., Sai Pravallika, M., & Kumar, S. (2019). Variability and time series trend analysis of rainfall over Krishna District of Andhra Pradesh: A case study. *International Journal of Recent Technology and Engineering*, *7*(6C2), 720–726.

Salehi, S., Dehghani, M., Mortazavi, S. M., & Singh, V. P. (2020). Trend analysis and change point detection of seasonal and annual precipitation in Iran. *International Journal of Climatology*, *40*(1), 308–323. https://doi.org/10.1002/joc.6211

Sasane, S. (2016). Detection of changes in rainfall for identifying climate change using Mann-Kendall test. *International Journal of Researches in Biosciences and Agriculture Technology*, *4*(1), 150–152. https://doi.org/10.29369/ijrbat.2016.04.i.0043

Sinha, A., Kumari, A., Mahapatra, S., Singh, H. P., & Bharti, B. (2019). Temporal rainfall variability and its correlation with temperature over Ranchi, Jharkhand. *International Journal of Engineering and Advanced Technology*, *9*(2), 1099–1104. https://doi.org/10.35940/ijeat.b3429.129219

Sridhar, S., & Raviraj, A. (2017). Statistical trend analysis of rainfall in Amaravathi River Basin using Mann-Kendall test. *Current World Environment*, *12*(1), 89–96. https://doi.org/10.12944/cwe.12.1.11

Sulaiman, N. H., Kamarudin, M. K. A., Mustafa, A. D., Amran, M. A., Azaman, F., Abidin, I. Z., & Hairoma, N. (2015). Analisis corak sungai Pahang menggunakan kaedah bukan parametrik: Ujian corak Mann Kendall. *Malaysian Journal of Analytical Sciences*, *19*(6), 1327–1334.

Tabari, H., Marofi, S., Aeini, A., Hosseinzadeh-Talaee, P., & Mohammadi, K. (2011). Trend analysis of reference evapotranspiration in the western half of Iran. *Agricultural and Forest Meteorology*, *151*(2), 128–136. https://doi.org/10.1016/j.agrformet.2010.09.009

Yilmaz, B. (2019). Analysis of hydrological drought trends in the GAP region (southeastern Turkey) by Mann-Kendall test and innovative Sen method. *Applied Ecology and Environmental Research*, *17*(2), 3325–3342. https://doi.org/10.15666/aeer/1702_33253342

Yue, S., Pilon, P., & Cavadias, G. (2002). Power of the Man-Kendall and Spearman's rho tests for detecting monotonic trends in hydrological series. *Journal of Hydrology*, *259*, 254–271. https://www.elsevier.com/locate/jhydrol

Chapter 11

Advancing Smart Agriculture in India: AI and IoT-driven Marketing Strategies for Sustainable Development

Arpan Anand[a] *and Piyush Gupta*[b]

[a]*Jaipuria Institute of Management, Noida, India*
[b]*Birla Global University, Bhubaneshwar, India*

Abstract

This research investigates the integration of artificial intelligence (AI) and the Internet of Things (IoT) within agricultural marketing strategies in the Indian context, employing a qualitative research methodology to construct a comprehensive theoretical framework. Through extensive case studies, phenomenological inquiry, and a grounded theory approach, we analyze how these technological innovations are reshaping agricultural practices and marketing paradigms. Our study contributes to the existing literature by proposing a novel theoretical framework that bridges the gap between technological adoption and sustainable agricultural development. By employing institutional theory, socio-technical systems theory, and the technology acceptance model (TAM), we provide nuanced insights into the complex interplay between technological innovation, institutional frameworks, and farmer adoption behaviors. This research enhances our understanding of the challenges and opportunities associated with the implementation of smart agriculture, offering both theoretical contributions and practical implications for policymakers and practitioners in the field. Ultimately, our findings underscore the necessity for a multidimensional approach to technology adoption that considers not only technological factors but also the socio-economic and

Sustainable and Smart Farming Eco-systems:
Changing Landscapes in the Age of Technology, 147–164

doi:10.1108/978-1-83608-092-320261011

institutional contexts that influence farmer engagement and sustainability outcomes.

Keywords: Smart agriculture; artificial intelligence; internet of things; agricultural marketing; sustainable development; institutional theory

1. Introduction

1.1. Background and Theoretical Context

The agricultural sector in India is at a critical juncture, where traditional practices are increasingly intersecting with technological innovations, particularly through the integration of AI and the IoT. This convergence signifies not only technological advancement but also a fundamental shift in the socio-technical landscape of agriculture (Antony et al., 2020; Swathi Kumari, 2023 ; Agrawal & Gupta, 2021). The adoption of these technologies has the potential to revolutionize agricultural practices, enhance productivity, and promote sustainability, thereby addressing pressing challenges such as food security and resource management (Antony et al., 2020; Ennouri et al., 2021). Recent studies have underscored the transformative potential of smart farming technologies in reshaping agricultural practices (Li et al., 2022; Özdoğan et al., 2017). However, the adoption of these technologies is not merely a technical challenge; it is a complex socio-cultural process deeply embedded within existing institutional frameworks (Sabrina et al., 2022). As Bronson (2018) articulates, the digital agricultural revolution must be contextualized within broader rural sociological transformations, highlighting the need for a multifaceted analytical approach. This research is situated at the intersection of several theoretical frameworks that provide a robust foundation for understanding the dynamics of technological integration in agriculture:

1. *Institutional theory*: This framework examines how formal and informal institutions influence technological adoption in agriculture, shaping farmers' decisions and practices. Recent work by Lawrence et al. (2022) emphasizes the role of institutional entrepreneurship in driving agricultural innovation.
2. *Socio-technical systems theory*: This perspective facilitates an understanding of the complex interplay between technological innovations and social structures within agricultural communities (Albarrak et al., 2022). The multi-level perspective (MLP) offered by Geels and Schot (2007) provides a valuable framework for analyzing technological transitions in agriculture.
3. *TAM*: This model analyzes the factors influencing farmer acceptance and adoption of new technologies (Chen et al., 2014). Recent extensions of TAM by Liu et al. (2020) have incorporated cultural and contextual factors specific to agricultural settings, enhancing its applicability in this domain. These theoretical underpinnings allow for a nuanced exploration of the qualitative dimensions of technological innovation in agriculture. As Wolfert et al. (2017) note, the convergence of these perspectives enables a deeper understanding of the challenges associated with smart farming adoption.

1.2. Research Objectives and Questions

This study aims to address the following research questions:

1. How do institutional frameworks shape the adoption of AI and IoT technologies in Indian agriculture? – What roles do formal and informal institutions play in facilitating or hindering technological adoption? – How do institutional logics evolve in response to technological innovations?
2. What role do social and cultural factors play in the acceptance or resistance to technological innovations in agricultural marketing? – How do community networks influence technology adoption decisions? – What cultural factors mediate the relationship between technological capability and actual usage?
3. How do farmers perceive and make sense of technological changes in their traditional practices? – What cognitive frameworks do farmers employ to understand and evaluate new technologies? – How do collective sensemaking processes influence adoption behaviors?
4. What are the implications of technological adoption for sustainable agricultural development? – How do AI and IoT technologies contribute to or potentially hinder sustainable farming practices? – What are the socio-economic impacts of technological adoption on rural communities?

1.3. Theoretical Significance

This research contributes significantly to the theoretical landscape by proposing an integrated framework combining institutional theory, socio-technical systems theory, and the extended TAM to explain technology adoption in Indian agriculture. By contextualizing AI and IoT adoption within India's socio-cultural and institutional contexts, the study offers rich qualitative insights into how traditional and modern logics interact to influence farmers' decisions. It highlights the evolution of hybrid practices and institutional arrangements that bridge cultural values and modern innovations, emphasizing the socio-economic impacts and sustainability outcomes of these transitions. This approach responds to calls for more nuanced, qualitative research in agricultural technology studies, offering practical and theoretical advancements.

2. Literature Review

2.1. Institutional Perspective on Agricultural Innovation

The adoption of technological innovations in agriculture is significantly influenced by institutional frameworks, which serve as the underlying structures that shape the behavior of individuals and organizations within the agricultural sector (Maffezzoli et al., 2022; Swathi Kumari, 2023). Institutional theory provides a valuable lens for understanding how both formal and informal institutions impact technological adoption. Recent work by Klerkx and Rose (2020) emphasizes the necessity of an institutional perspective in comprehending agricultural innovation systems (Dey & Singh, 2020). This section explores the various dimensions of institutional influence on agricultural innovation:

1. *Regulatory institutions*: Government policies, legal frameworks, and regulatory environments play a crucial role in shaping technological adoption (Ennouri et al., 2021). Recent studies by Lowndes and Roberts (2013) have highlighted how regulatory frameworks can both enable and constrain technological innovation in agriculture. For instance, supportive policies can incentivize the adoption of sustainable practices, while overly stringent regulations may hinder innovation.
2. *Normative institutions*: Social norms, values, and expectations within agricultural communities significantly influence farmers' decisions to adopt new technologies (Antony et al., 2020). Research by Thornton et al. (2012) demonstrates how normative institutions can either promote or inhibit the adoption of sustainable technologies, depending on the prevailing cultural attitudes toward innovation and change.
3. *Cultural-cognitive institutions*: Shared understandings and cultural frameworks shape how technology is perceived and utilized in agricultural settings (Swathi Kumari, 2023). Scholars such as Klerkx et al. (2021) have emphasized the importance of cultural-cognitive elements in shaping farmers' responses to digital technologies. For example, farmers' beliefs about the efficacy and reliability of new technologies can significantly impact their willingness to adopt them. Recent studies have also emphasized the role of institutional logics in shaping technological adoption in agriculture (Li et al., 2022; Özdoğan et al., 2017). For instance, Jha and Singh (2016) found that traditional farming practices are deeply embedded in cultural institutions, creating potential resistance to technological innovations. This perspective has been further developed by scholars such as Lowenberg-DeBoer et al. (2019), who examine how competing institutional logics influence the adoption of precision agriculture technologies.

2.1.1. Institutional Entrepreneurship in Agriculture

The concept of institutional entrepreneurship (DiMaggio, 1988) has gained traction in agricultural studies, highlighting how certain actors can drive change within institutional frameworks (Sabrina et al., 2022). Recent work by Hermans et al. (2015) identifies key actors in the agricultural sector who act as institutional entrepreneurs, creating new practices and norms around technological innovation. Key findings include:

1. *Boundary-spanning organizations*: These organizations play a critical role in facilitating technological adoption by bridging gaps between different stakeholders and fostering collaboration.
2. *Legitimacy-building strategies*: Institutional entrepreneurs often engage in legitimacy-building strategies to introduce new technologies, ensuring that innovations are accepted and integrated into existing practices (Albarrak et al., 2022).

3. *New organizational forms*: The development of new organizational structures, such as cooperatives and networks, supports technological innovation by providing resources and collective learning opportunities (Chen et al., 2014).

2.1.2. Institutional Complexity and Hybrid Arrangements

Agricultural technology adoption often occurs in contexts characterized by institutional complexity, where multiple, sometimes conflicting, institutional logics coexist (Li et al., 2022). Recent work by Klerkx et al. (2021) has examined how farmers navigate this complexity:

1. *Hybrid practices*: Farmers often develop hybrid practices that combine traditional and modern approaches, allowing them to adapt to changing circumstances while maintaining cultural relevance (Zakerabasali & Ayyoubzadeh, 2022).
2. *New institutional arrangements*: The creation of new institutional arrangements, such as partnerships between public and private sectors, supports technological adoption by providing necessary resources and expertise (Sood et al., 2021).
3. *Evolution of existing institutions*: Existing institutions may evolve to accommodate technological change, reflecting the dynamic nature of agricultural innovation systems (Swathi Kumari, 2023).

2.2. Socio-technical Systems in Agriculture

The socio-technical systems' perspective provides valuable insights into the complex interrelationships between technological innovations and social structures in agriculture (Adli et al., 2023; Chen et al., 2021). Recent work by Klerkx and Rose (2020) emphasizes the need for a systems thinking approach to understanding agricultural innovation, recognizing that technology does not exist in isolation but is embedded within broader socio-economic contexts.

2.2.1. Multi-level Perspective

The MLP framework offers a structured approach to understanding transitions in agricultural systems (Brocke et al., 2015). This framework identifies three levels of analysis:

1. *Niche innovations*: These are experimental spaces where new technologies are developed and tested. Recent work by Pigford et al. (2018) examines how protected spaces, such as living labs and experimental farms, facilitate the development of agricultural innovations (Batra & Keller, 2016).
2. *Socio-technical regimes*: Established practices, rules, and technologies provide stability within agricultural systems. Scholars such as Elzen et al. (2012) have identified key elements of agricultural regimes, including knowledge and capabilities, sectoral policies, and market structures.

3. *Landscape developments*: Broader contextual factors, such as climate change and global food security concerns, influence the possibilities for transition within agricultural systems. Recent research by Herrero et al. (2020) highlights the significance of these landscape factors in shaping agricultural innovation. Recent research by Kumar et al. (2019) demonstrates how the MLP framework can explain the successful adoption of precision agriculture technologies in certain Indian states while highlighting barriers in others. This work has been extended by scholars such as Rose et al. (2016), who emphasize the role of intermediaries in facilitating transitions between levels.

2.2.2. Strategic Niche Management (SNM)

SNM focuses on how technological innovations can be nurtured in protected spaces (Mohr & Kühl, 2021). In the agricultural context, SNM has been applied through various mechanisms:

1. *Demonstration farms*: These serve as living laboratories for new technologies, facilitating peer-to-peer learning and adaptation to local contexts (Sabrina et al., 2022). Research by Eastwood et al. (2017) shows how demonstration farms can mitigate risks associated with new technology adoption.
2. *Farmer producer organizations* (FPOs): FPOs create collective learning environments that pool resources for technology adoption, collective bargaining, and knowledge sharing (Victoire, 2023). Recent work by Trebbin (2014) highlights the role of FPOs in enhancing farmers' capacity to adopt new technologies.
3. *Public–private partnerships* (PPPs): These partnerships provide essential resources and institutional support for technological innovation. Studies by Schut et al. (2016) demonstrate how PPPs can bridge funding gaps and facilitate knowledge transfer between sectors.

2.3. Technology Acceptance in Agricultural Communities

Building on the TAM (Davis, 1989), recent scholarly work has explored the unique factors influencing farmer adoption behaviors.

2.3.1. Extended TAM for Agriculture

Our research extends the traditional TAM by incorporating additional dimensions relevant to agricultural contexts:

1. *Cultural factors*: Cultural norms significantly influence technology perception and acceptance. Recent work by Liu et al. (2020) identifies the role of traditional knowledge systems and community hierarchies in shaping farmers' attitudes toward technology (Mariani et al., 2021).
2. *Collective Decision-making*: The role of community dynamics in adoption decions is critical. Studies by Rose et al. (2016) highlight the importance

of community leaders and agricultural extension workers in facilitating technology adoption (Uzhinskiy, 2023).

3. *Risk perception*: Farmers' evaluations of technological risks play a crucial role in their adoption decisions. Research by Crane et al. (2018) examines how financial, social, and environmental risks influence farmers' willingness to adopt new technologies (Nalbant & Aydin, 2023).

2.3.2. Empirical Evidence

Recent studies have provided valuable insights into technology acceptance in agriculture. Rose et al. (2016) found that trust in technology providers significantly influences adoption decisions, emphasizing the importance of local partnerships and demonstration effects (Barrile et al., 2022). Li et al. (2022) highlighted the role of demonstration effects in facilitating technology adoption, noting that peer learning networks and visual evidence of technology benefits are critical.

Our research builds on these findings while addressing their limitations by incorporating institutional perspectives, examining long-term adoption patterns, and considering socio-economic factors (Eastwood et al., 2017; Barrett et al., 2020).

3. Methodology

3.1. Research Philosophy and Approach

This study is grounded in an interpretive research philosophy, which emphasizes the need to explore and understand technological adoption in agriculture through the subjective experiences of farmers. This approach is particularly suited to capturing the complex interplay between technological integration and socio-cultural factors that shape adoption behaviors. Drawing on recent calls in the literature for more nuanced, context-sensitive research methodologies in agricultural studies (Maffezzoli et al., 2022; Swathi Kumari, 2023), this research prioritizes a detailed examination of farmers' perspectives, highlighting the influence of local contexts and individual experiences.

3.1.1. Qualitative Approach Rationale

To comprehensively understand the phenomena under investigation, this study employs a qualitative research methodology that integrates multiple complementary approaches.

Phenomenology forms the foundation of this inquiry, focusing on the lived experiences of farmers as they encounter and adapt to technological changes. Through individual narratives, the study illuminates the emotional, psychological, and cognitive dimensions of technology adoption. By delving into farmers' meaning-making processes, this approach captures how they interpret and derive significance from their interactions with AI and IoT technologies in agriculture (Ennouri et al., 2021; Van Manen, 2016).

Grounded theory further enriches the methodological framework by allowing theoretical insights to emerge directly from the qualitative data. This involves iterative coding and analysis, ensuring that emerging theories are refined through a cyclical process of data collection and interpretation. Theoretical sampling ensures that participants are selected for their relevance to the evolving framework, while the constant comparative method systematically identifies patterns and themes across data segments (Antony et al., 2020; Charmaz, 2006).

The case study approach complements these methods by providing a rich, contextual understanding of specific instances of technology adoption. By employing a multiple-case design, the study enhances the generalizability of findings (Yin, 2018). Embedded units of analysis capture diverse perspectives within each case, while cross-case synthesis enables the identification of commonalities and differences, offering deeper insights into the dynamics of agricultural innovation (Özdoğan et al., 2017).

3.2. Data Collection Methods

To ensure robust and reliable findings, the study employs a multi-method data collection strategy, enabling triangulation and the generation of rich data.

3.2.1. In-depth Interviews

A total of 65 semi-structured interviews were conducted, categorized as follows:

- *Farmers* (40 interviews): Participants were stratified based on farm size (small, medium, and large), level of technological adoption (early adopters and resistant individuals), and geographical location, ensuring representation across regional variations in agricultural practices.
- *Agricultural technology providers* (15 interviews): This group included representatives from multinational corporations (offering perspectives on large-scale technology supply), local startups (providing insights into innovative, localized solutions), and agricultural extension services (highlighting their role in facilitating technology adoption).
- *Policymakers* (10 interviews): These participants comprised government officials, agricultural policy experts, and nongovernmental organization (NGO) representatives, offering strategic insights into policy frameworks and grassroots implementation.

The interview protocols were refined iteratively based on emerging themes, ensuring alignment with best practices for qualitative research design (Patton, 2015).

3.2.2. Focus Group Discussions (FGDs)

FGDs were conducted to gather collective insights from diverse stakeholders. Eight focus groups, each comprising 6–8 participants, were organized to represent varied demographics, including age, gender, farm size, and technological expertise. These groups included farmers, technology providers, and policymakers, ensuring balanced representation and varied perspectives.

Discussions were structured to last 90–120 minutes, employing methods such as audio recording for accuracy, facilitator notes for non-verbal observations, and visual aids like diagrams and charts to stimulate engagement. This approach facilitated rich, interactive dialogues that captured a spectrum of experiences and viewpoints.

3.2.3. Ethnographic Observation

Extended field visits to five agricultural communities allowed for in-depth ethnographic observation. Researchers engaged in participant observation, immersing themselves in daily farming routines, community meetings, and technology training sessions. Field notes and photographic documentation provided detailed records of interactions and technology use. Informal conversations with community members helped uncover unofficial narratives and social dynamics, revealing hidden barriers to technology adoption.

3.3. Data Analysis

The data analysis process adhered to rigorous qualitative research practices, employing thematic analysis and structured coding techniques to derive insights.

3.3.1. Thematic Analysis

Thematic analysis followed Braun and Clarke's (2006) six-step process, which included familiarizing with the data, generating initial codes using NVivo software, identifying patterns, refining themes through peer review and participant feedback, defining and naming themes, and integrating findings into the theoretical framework for reporting.

3.3.2. Coding Process

The coding process involved three stages: open coding (line-by-line identification of initial concepts), axial coding (establishing relationships between concepts and developing categories), and selective coding (refining the theoretical framework and validating it against the data). This iterative process ensured a robust analytical approach, balancing description with interpretation.

3.4. Research Quality and Ethical Considerations

To ensure the credibility, dependability, transferability, and confirmability of the findings, the study adhered to Lincoln and Guba's (1985) criteria. Strategies included member checking, prolonged engagement, audit trails, and reflexivity.

3.4.1. Ethical Considerations

Strict ethical guidelines were followed throughout the research. Informed consent was obtained from all participants, emphasizing voluntary participation and the right to withdraw at any stage. Confidentiality and anonymity were maintained through pseudonyms and secure data storage. Data security measures, including encrypted storage and limited access, were implemented to protect participant information.

This comprehensive methodology ensures that the research is ethically sound, contextually sensitive, and capable of yielding rich insights into the adoption of AI and IoT technologies in Indian agriculture.

4. Findings

4.1. Institutional Logics and Technology Adoption

The findings of this study reveal a complex interplay between institutional logics and the adoption of technology in agriculture. This interplay underscores the tension between traditional farming practices and the influence of modern technological innovations. Farmers find themselves navigating this dynamic landscape, balancing long-established methods with the demands of new technologies (refer Table 11.1).

4.1.1. Traditional Versus Modern Logics

Farmers' decision-making processes reflect a duality between reliance on ancestral knowledge and the potential offered by data-driven approaches. Many farmers exhibit skepticism toward data-driven methods, preferring time-tested

Table 11.1. Derived Themes.

Theme	Description	Key Insights
Institutional logics	The interaction of traditional and modern practices in decision-making and risk assessment.	Farmers combine ancestral knowledge with technological inputs, reflecting a hybrid adaptation approach.
Legitimacy and trust	Building trust through intermediaries and local success stories.	Demonstrations and peer networks enhance credibility and willingness to adopt.
Socio-technical transitions	Mechanisms like demonstration farms and market demands drive technology adoption.	Early adopter networks and policies create an enabling environment.
Collective sensemaking	Community-based learning and intergenerational knowledge transfer facilitate adaptation.	Younger farmers often act as mediators, ensuring technology aligns with cultural practices.

Source: Authors' compilation.

practices rooted in tradition. This tension extends to the roles of community elders and technical experts, where farmers often seek guidance from trusted local figures, sometimes sidelining the expertise of technology providers. Furthermore, traditional practices grounded in seasonal rhythms often conflict with real-time monitoring systems promoted by modern technologies, creating operational and strategic challenges for farmers.

Risk assessment strategies also illustrate this tension. Traditional methods emphasize community-based practices for mitigating risks, fostering a sense of collective responsibility. In contrast, technological solutions tend to focus on individualized approaches, shifting responsibility to the farmer. Similarly, traditional knowledge systems offer holistic perspectives, integrating various aspects of farming, while modern technologies frequently emphasize specialized solutions.

4.1.2. Legitimacy and Trust

The adoption of new technologies is significantly influenced by the legitimacy conferred through trusted institutional actors. Agricultural extension workers, for instance, serve as critical intermediaries, bridging the gap between technology providers and farmers. Local success stories within farming communities further enhance the credibility of these innovations, establishing trust through relatable examples.

Mechanisms for building trust include peer-to-peer learning networks, which facilitate knowledge sharing and create a supportive environment for hesitant adopters. Demonstration effects – where farmers witness the successful application of technologies by their peers – play a vital role in encouraging adoption. Additionally, a gradual introduction of technologies allows farmers to build confidence over time, fostering long-term engagement.

4.2. Socio-technical Transitions

The study identifies three mechanisms of socio-technical transitions that enable the adoption of agricultural technologies, highlighting the role of protected experimentation spaces, external pressures, and evolving aspirations.

4.2.1. Niche Development

Protected spaces such as demonstration farms provide farmers with environments to safely experiment with new technologies without risking financial losses. These farms act as knowledge hubs, disseminating best practices and facilitating the adaptation of technologies to local contexts. Early adopter networks also play a crucial role by forming support systems, sharing strategies for overcoming challenges, and creating local expertise that guides hesitant farmers.

Institutional support further strengthens these niches through policy frameworks, financial incentives, and technical assistance. Supportive government policies and NGO initiatives create an enabling environment for technology adoption, while subsidies and training programs reduce the barriers associated with cost and knowledge gaps.

4.2.2. Landscape Pressures

External pressures, such as climate change and market demands for traceability, create opportunities for technological transitions in agriculture. Climate change disrupts traditional farming practices, prompting farmers to adopt resilient strategies through modern technologies. Similarly, growing consumer interest in sustainably produced food and regulatory requirements around traceability compel farmers to integrate technologies that improve market access and enhance profitability.

The aspirations of younger farmers also drive technological adoption. More educated and technologically savvy, younger generations exhibit greater openness to innovation. Economic pressures further motivate efficiency improvements, while the desire to modernize aligns with their values and ambitions, encouraging the use of advanced farming practices.

4.3. Technology Sensemaking and Adaptation

Farmers often engage in collective sensemaking processes to understand and adapt to new technologies. Community spaces, both formal and informal, serve as vital venues for sharing knowledge and solving problems collectively. Regular gatherings facilitate knowledge-sharing rituals that foster a collaborative learning environment.

Intergenerational knowledge transfer also plays a pivotal role, with younger farmers acting as mediators who bridge the gap between traditional practices and modern technologies. By translating technical information into accessible formats, they enable their communities to engage effectively with new tools and approaches. Adapting communication styles ensures that all members, regardless of their technological proficiency, can participate in these discussions and decisions.

4.4. Summary

In conclusion, the findings emphasize the intricate dynamics of institutional logics, socio-technical transitions, and collective sensemaking processes in shaping the adoption of AI and IoT technologies in Indian agriculture. These insights underscore the importance of addressing contextual factors that influence farmers' adoption behaviors and the need for tailored approaches to facilitate successful integration of technologies. The results also highlight the transformative potential of collaboration, trust-building, and adaptive practices in driving sustainable agricultural development.

5. Discussion

5.1. Theoretical Contributions

5.1.1. Extended Institutional Theory

This study advances the understanding of technological adoption in agriculture by proposing a model of "adaptive institutional logics." This framework explicates the mechanisms through which actors navigate the complexities of integrating technology into traditional agricultural practices. A key dimension of this model is the *blending of competing logics*, wherein farmers combine traditional knowledge with modern technological practices. This hybridization respects cultural heritage while embracing the potential of innovation, leading to the creation of new organizational forms. Such structures, including cooperatives and farmer organizations, facilitate technology adoption by bridging traditional methods with modern strategies. Decision-making processes also evolve under this dynamic, as farmers increasingly balance ancestral wisdom with data-driven insights to make more informed and contextually relevant choices.

Another critical dimension involves the *creation of legitimacy bridges* to foster acceptance of new technologies. Boundary spanners – individuals or organizations that mediate between traditional practices and modern innovations – play a pivotal role in building trust and facilitating integration. Cultural alignment emerges as a significant factor, as technologies that resonate with local values are more readily embraced. Legitimacy is often established gradually, requiring sustained efforts such as community engagement and demonstrative benefits to ensure widespread acceptance.

Finally, this framework emphasizes the *development of hybrid institutional forms*, reflecting the adaptive nature of institutional frameworks in agriculture. These adaptations manifest in modified governance mechanisms and evolving social norms that collectively navigate the integration of traditional and modern practices.

5.1.2. Socio-technical Transitions in Traditional Contexts

The study introduces a "cultural alignment model" to explain technological transitions, particularly in traditional agricultural contexts. This model underscores the importance of *cultural resonance*, highlighting that successful adoption depends on aligning new practices with the values and beliefs of the farming community. Technologies must not only respect traditional knowledge systems but also adapt to local agricultural practices to ensure relevance and effectiveness.

Social learning mechanisms are equally vital, with collective learning environments fostering knowledge sharing and innovation. Younger farmers often act as mediators, bridging traditional and modern knowledge systems to cultivate a culture of adaptation and progress. Joint experimentation further reinforces community bonds, as farmers collaboratively explore and refine the use of new technologies.

5.2. Practical Implications

5.2.1. Policymakers

The findings underscore the need for *culturally sensitive technology policies* that integrate local knowledge systems into interventions. Policymakers should prioritize support for hybrid practices, facilitating the coexistence of traditional and modern methods. Flexible implementation frameworks are essential to address the diverse needs of farming communities effectively.

Funding for research on adaptive practices should be increased, focusing on the integration of traditional and modern approaches. Recognizing informal knowledge networks can enhance information dissemination and support farmer-driven innovation initiatives. Furthermore, institutional intermediaries, such as agricultural extension workers and boundary-spanning organizations, play a crucial role in facilitating adoption and building trust within communities. Supporting farmer-to-farmer networks can also promote peer learning and community resilience.

5.2.2. Technology Providers

Technology providers should prioritize *designing adaptable solutions* that cater to the varied needs of farmers. Modular systems and user-friendly interfaces can improve accessibility for individuals with different levels of technological proficiency. Integrating technologies with existing practices minimizes disruptions and enhances acceptance.

Engagement with *institutional gatekeepers*, including community leaders and local institutions, is essential for building credibility and ensuring relevance. Collaborative efforts with stakeholders help align technologies with community needs, fostering long-term trust and sustained engagement. Collective adoption strategies, such as community-based implementation and peer learning networks, facilitate knowledge sharing and reduce risks associated with new technology adoption.

5.3. Future Research Directions

The study identifies several avenues for future research. First, *longitudinal studies* can provide deeper insights into the long-term impacts of hybrid practices and the evolution of institutional frameworks in agriculture. Second, *comparative analysis across cultural contexts* can shed light on how diverse institutional environments influence technology adoption, enhancing the transferability of findings (Verma & Choudhary, 2023).

Lastly, investigating *power dynamics in technological transitions* – including the gender dimensions of adoption – can reveal inequities in access and benefits. Understanding how these transitions impact social hierarchies and economic divides can offer a more comprehensive view of the socio-economic implications of agricultural innovation.

6. Conclusion

This study contributes to the growing body of knowledge on AI and IoT adoption in traditional agricultural contexts. It highlights several critical aspects, including the importance of institutional logics, socio-technical systems, and local adaptation processes. The findings emphasize that competing logics, such as traditional and modern practices, are central to understanding farmers' navigation of technological adoption. Hybrid practices reflect the adaptive nature of institutional frameworks, while legitimacy-building efforts foster trust and acceptance.

The interplay between social and technical elements is crucial for successful technology integration. Protected niches provide safe environments for experimentation, while external pressures such as climate change and market demands create opportunities for innovation. Collective sensemaking processes enable farmers to collaboratively adapt technologies, ensuring that innovations resonate with local practices and values.

In an era where agriculture must balance sustainability and efficiency, this study offers theoretical insights and practical guidance for navigating these transitions. By emphasizing context, culture, and community engagement, it provides a nuanced understanding of the complexities surrounding technological adoption in agriculture.

References

Adli, H., Remli, M., Wong, K., Ismail, N., González-Briones, A., Corchado, J., & Mohamad, M. (2023). Recent advancements and challenges of AIoT application in smart agriculture: A review. *Sensors*, *23*(7), 3752. https://doi.org/10.3390/s23073752

Agrawal, A., & Gupta, R. (2021). The role of AI and IoT in transforming Indian agriculture: A review. *Journal of Agricultural Science and Technology*, *23*(4), 123–145.

Albarrak, K., Gulzar, Y., Hamid, Y., Mehmood, A., & Soomro, A. (2022). A deep learning-based model for date fruit classification. *Sustainability*, *14*(10), 6339. https://doi.org/10.3390/su14106339

Antony, A., Leith, K., Jolley, C., Lu, J., & Sweeney, D. (2020). A review of practice and implementation of the Internet of Things (IoT) for smallholder agriculture. *Sustainability*, *12*(9), 3750. https://doi.org/10.3390/su12093750

Barrett, M., et al. (2020). Contextualizing agricultural technology adoption: A review of the literature. *Agricultural Systems*, *182*, 102–111.

Barrile, V., Simonetti, S., Citroni, R., Fotia, A., & Bilotta, G. (2022). Experimenting agriculture 4.0 with sensors: A data fusion approach between remote sensing UAVs and self-driving tractors. *Sensors*, *22*(20), 7910. https://doi.org/10.3390/s22207910

Batra, R., & Keller, K. (2016). Integrating marketing communications: New findings, new lessons, and new ideas. *Journal of Marketing*, *80*(6), 122–145. https://doi.org/10.1509/jm.15.0419

Braun, V., & Clarke, V. (2006). Using thematic analysis in psychology. *Qualitative Research in Psychology*, *3*(2), 77–101. https://doi.org/10.1191/1478088706qp063oa

Brocke, J., Simons, A., Riemer, K., Niehaves, B., Plattfaut, R., & Cleven, A. (2015). Standing on the shoulders of giants: Challenges and recommendations of literature search in information systems research. *Communications of the Association for Information Systems*, *37*. https://doi.org/10.17705/1CAIS.03709

Bronson, K. (2018). Smart farming: Including rights holders for responsible agricultural innovation. *Technology Innovation Management Review*, *8*(2), 7–12.

Charmaz, K. (2006). *Constructing grounded theory: A practical guide through qualitative analysis*. Sage Publications.

Chen, J., Hu, T., Wu, J., Ping, H., & Lin, K. (2014). Applications of internet of things in facility agriculture. *Applied Mechanics and Materials*, *685*, 517–523. https://doi.org/10.4028/www.scientific.net/amm.685.517

Chen, L., Jiang, M., Jia, F., & Liu, G. (2021). Artificial intelligence adoption in business-to-business marketing: Toward a conceptual framework. *Journal of Business and Industrial Marketing*, *37*(5), 1025–1044. https://doi.org/10.1108/JBIM-09-2020-0448

Crane, A., Henriques, I., & Husted, B. (2018). Quo vadis, stakeholder theory? Past, present, and future. *Academy of Management Annals*, *12*(1), 149–177.

Davis, F. D. (1989). Perceived usefulness, perceived ease of use, and user acceptance of information technology. *MIS Quarterly*, *13*(3), 319–340. https://doi.org/10.2307/249008

Dey, A., & Singh, R. (2020). Understanding the adoption of smart farming technologies in India: An institutional perspective. *International Journal of Agricultural Management*, *9*(2), 78–92.

DiMaggio, P. J. (1988). Interest and agency in institutional theory. In L. Zucker (Ed.), *Institutional patterns and organizations: Culture and environment* (pp. 3–21). Ballinger.

Eastwood, C., Klerkx, L., Ayre, M., & Rue, B. (2017). Managing socio-ethical challenges in the development of smart farming: From a fragmented to a comprehensive approach for responsible research and innovation. *Journal of Agricultural and Environmental Ethics*, *32*(5–6), 741–768. https://doi.org/10.1007/s10806-017-9704-5

Elzen, B., Augustyn, A. M., Barbier, M., & van Mierlo, B. (Eds.). (2012). *AgroEcological transitions: Changes and breakthroughs in the making*. Wageningen University & Research.

Ennouri, K., Smaoui, S., Gharbi, Y., Cheffi, M., Braïek, O., Ennouri, M., & Triki, M. (2021). Usage of artificial intelligence and remote sensing as efficient devices to increase agricultural system yields. *Journal of Food Quality*, *2021*, 1–17. https://doi.org/10.1155/2021/6242288

Geels, F. W., & Schot, J. (2007). Typology of sociotechnical transition pathways. *Research Policy*, *36*(3), 399–417.

Herrero, M., Thornton, P. K., Power, B., Bogard, J. R., Remans, R., Fritz, S., Gerber, J. S., Nelson, G., See, L., Waha, K., Watson, R. A., West, P. C., Samberg, L. H., van de Steeg, J., Stephenson, E., van Wijk, M., & Havlík, P. (2020). Farming and the geography of nutrient production for human use: A transdisciplinary analysis. *Lancet Planetary Health*, *4*(8), 318–329.

Hermans, F., Klerkx, L., & Roep, D. (2015). Structural conditions for collaboration and learning in innovation systems: Expressed by farmers, researchers and advisors. *The Journal of Agricultural Education and Extension*, *21*(3), 203–219.

Jha, S., & Singh, T. (2016). Institutional innovations for agricultural development: A case study of FPOs in India. *Agricultural Economics Research Review*, *29*(3), 405–416.

Klerkx, L., Jakku, E., & Labarthe, P. (2019). A review of social science on digital agriculture, smart farming and agriculture 4.0: New contributions and a future research agenda. *NJAS – Wageningen Journal of Life Sciences*, *90–91*, 100315.

Kumar, A., Tiwari, P., & Singh, R. K. (2019). Multi-level perspective on precision agriculture adoption: Evidence from Indian states. *Journal of Rural Studies*, *67*, 220–231.

Lawrence, T. B., Suddaby, R., & Leca, B. (2022). Institutional work: Refocusing institutional studies of organization. *Journal of Management Inquiry*, *11*(1), 24–30.

Li, Y., Yang, J., & Chen, D. (2022). The intelligent path planning system of agricultural robots via reinforcement learning. *Sensors*, *22*(12), 4316. https://doi.org/10.3390/s22124316

Lincoln, Y. S., & Guba, E. G. (1985). *Naturalistic inquiry*. Sage Publications.

Liu, X., Jin, Z., & Zou, Z. (2020). Cultural influences on technology adoption: Evidence from Chinese and Western farmers. *Journal of International Agricultural Trade and Development*, *16*(4), 52–63.

Lowenberg-DeBoer, J., Griffin, T. W., & Lambert, D. M. (2019). Economics of precision agriculture. In Q. Zhang (Ed.), *Precision agriculture technology for crop farming* (pp. 255–279). Springer.

Lowndes, V., & Roberts, M. (2013). *Why institutions matter: The new institutionalism in political science*. Palgrave Macmillan.

Maffezzoli, F., Ardolino, M., Bacchetti, A., Perona, M., & Renga, F. (2022). Agriculture 4.0: A systematic literature review on the paradigm, technologies, and benefits. *Futures*, *142*, 102998. https://doi.org/10.1016/j.futures.2022.102998

Mariani, M., Perez-Vega, R., & Wirtz, J. (2022). AI in marketing, consumer research and psychology: A systematic literature review and research agenda. *Psychology & Marketing*, *39*(4), 755–776.

Mohr, S., & Kühl, R. (2021). Acceptance of artificial intelligence in German agriculture: An application of the Technology Acceptance Model and the Theory of Planned Behavior. *Precision Agriculture*, *22*(6), 1816–1844. https://doi.org/10.1007/s11119-021-09814-x

Özdoğan, B., Gacar, A., & Aktaş, H. (2017). Digital agriculture practices in the context of agriculture 4.0. *Pressacademia*, *4*(2), 184-191. https://doi.org/10.17261/pressacademia.2017.448

Patton, M. Q. (2015). *Qualitative research & evaluation methods: Integrating theory and practice* (4th ed.). Sage Publications.

Pigford, A. A. E., Hickey, G. M., & Klerkx, L. (2018). Beyond agricultural innovation systems? Exploring an agricultural innovation ecosystems approach for niche design and development in sustainability transitions. *Agricultural Systems*, *164*, 116–121. https://doi.org/10.1016/j.agsy.2018.04.007

Rose, D. C., Sutherland, W. J., Parker, C., Lobley, M., Winter, M., Morris, C., Twining, S., Ffoulkes, C., Amano, T., & Dicks, L. V. (2016). Decision support tools for agriculture: Towards effective design and delivery. *Agricultural Systems*, *149*, 165–174.

Sabrina, F., Sohail, S., Farid, F., Jahan, S., Ahamed, F., & Gordon, S. (2022). An interpretable artificial intelligence-based smart agriculture system. *Computers, Materials & Continua*, *72*(2), 3777–3797. https://doi.org/10.32604/cmc.2022.026363

Schut, M., Klerkx, L., Sartas, M., Lamers, D., Campbell, M. M. C., Ogbonna, I., Kaushik, M., Adekunle, A., & Leeuwis, C. (2016). Innovation platforms: Experiences with their institutional embedding in agricultural research for development. *Experimental Agriculture*, *52*(4), 537–561. https://doi.org/10.1017/S001447971500023X

Sood, A., Sharma, R., & Bhardwaj, A. (2021). Artificial intelligence research in agriculture: A review. *Online Information Review*, *46*(6), 1054–1075. https://doi.org/10.1108/oir-10-2020-0448

Swathi Kumari, S. (2023). Revolutionizing agriculture: A case study of IBM's AI innovations. *International Journal of Applied Engineering and Management Letters*, 95–114. https://doi.org/10.47992/ijaeml.2581.7000.0195

Thornton, P. K., Jones, P. G., Ericksen, P. J., & Challinor, A. J. (2011). Agriculture and food systems in sub-Saharan Africa in a 4°C world. *Philosophical Transactions of the Royal Society A*, *369*(1934), 117–136.

Trebbin, A. (2014). Farmer producer companies in India: A new concept for collective action? *Environment and Development Economics*, *19*(2), 207–232.

Uzhinskiy, A. (2023). Advanced technologies and artificial intelligence in agriculture. *Appliedmath*, *3*(4), 799–813. https://doi.org/10.3390/appliedmath3040043

Van Manen, M. (2016). *Researching lived experience: Human science for an action sensitive pedagogy*. Althouse Press.

Verma, R., & Choudhary, A. (2023). Challenges and opportunities in smart agriculture: Insights from Indian farmers. *Agricultural Systems*, *198*, 103–115.

Victoire, D. (2023). Leveraging artificial intelligence for enhancing agricultural productivity and sustainability. *Quing International Journal of Innovative Research in Science and Engineering*, *2*(2), 141–156. https://doi.org/10.54368/qijirse.2.2.0016

Wolfert, S., Ge, L., Verdouw, C., & Bogaardt, M. J. (2017). Big data in smart farming: A review. *Agricultural Systems*, *153*, 69–80. https://doi.org/10.1016/j.agsy.2017.01.023

Yin, R. K. (2018). *Case study research and applications: Design and methods*. Sage Publications.

Zakerabasali, S., & Ayyoubzadeh, S. (2022). Internet of things and healthcare system: A systematic review of ethical issues. *Health Science Reports*, *5*(6). https://doi.org/10.1002/hsr2.863

Chapter 12

Examining the Sustainability of Blockchain-driven Smart Agriculture Systems: A Triple Bottom Line Perspective

Harsh Shukla[a], *Vartul Shrivastava*[b] *and Shekhar Shukla*[c]

[a]*Department of Data Science and Engineering, Indian Institute of Science Education and Research Bhopal, Bhopal, India*
[b]*Information Systems Area, Indian Institute of Management Bangalore, Bengaluru, India*
[c]*Information Systems Area, Indian Institute of Management Indore, Indore, India*

Abstract

This chapter aims to assess the sustainability of blockchain-driven smart agriculture (BDSA) systems through the lens of the triple bottom line (TBL) framework, which integrates social, environmental, and economic aspects. We aim to provide a comprehensive examination of the impact, challenges, limitations, and future research directions of BDSA systems. This chapter evaluates the potential intersection of smart agriculture with blockchain technology. By synthesizing insights from existing literature on emerging technologies in smart agriculture, we have summarized and highlighted the key aspects of BDSA. We have employed the TBL framework to synthesize and organize findings, as it emphasizes sustainability through its social, environmental, and economic dimensions. The findings of this study highlight the merits and key challenges of BDSA, as well as specific issues related to technical awareness gaps, high energy consumption, and hardware compatibility. BDSA can promote sustainability by enhancing resource management, reducing waste, minimizing costs, and fostering trust among relevant stakeholders. This study can help stakeholders and agricultural corporations determine their future course of action. As blockchain can enhance transparency, BDSA might streamline supply chains and facilitate direct market access for concerned professionals. However, BDSA can face a range of challenges due to blockchain, including

Sustainable and Smart Farming Eco-systems:
Changing Landscapes in the Age of Technology, 165–181

doi:10.1108/978-1-83608-092-320261012

high implementation costs and varied regulatory frameworks across nations. This study provides a comprehensive analysis of the convergence of blockchain with smart agriculture, drawing on existing literature. Thus, it bridges the knowledge gap by explicating relevant insights for concerned readers.

Keywords: Smart agriculture; blockchain technology; triple bottom line (TBL); sustainability; supply chain management; transparency

1. Introduction

The constant supply of food worldwide largely depends on farming activities. The entire cultivation process, from planting to harvesting and distribution of the harvested crops, involves interactions among multiple direct and indirect stakeholders. Direct stakeholders include farmers and sellers of agricultural products, while indirect stakeholders include various consumer strata. The farm process involves two segments: pre-harvest and post-harvest. In the pre-harvest segment, the farming process encompasses cultivation (Jellason et al., 2021), while the post-harvest segment includes distribution and open market consumption of produce (Pranto et al., 2021).

The global population has increased rapidly in recent years. To meet this evolving demand, agricultural practitioners are looking to adopt new technologies in their cultivation practices. The use of innovative technologies to enhance farming practices is referred to as smart agriculture. These innovative farming practices can improve productivity and sustainability. Typically, traditional farming practices often face hurdles related to data falsification, traceability, and inefficient cost management. These hurdles can be mitigated by integrating emerging technologies, such as blockchain and internet of things (IoT), into farming practices. Contemporary blockchain technology entered mainstream discourse following Nakamoto's Bitcoin design, which described a peer-to-peer system for recording transactions without reliance on a central intermediary (Laurence, 2019). Notably, it emphasized direct online transactions from one source to another without relying on intermediaries (Sarmah, 2018). Considering these characteristics, blockchain could help mitigate the hurdles to the sustainability of traditional farming practices. Therefore, in this study, we analyze the sustainability of BDSA systems using the TBL framework to understand how blockchain can contribute across social, environmental, and economic dimensions. We use a TBL framework to systematically assess tenets of sustainability in BDSA, as it incorporates three significant pillars: social, environmental, and economic (Slaper & Hall, 2011).

From a social sustainability perspective, blockchain can help minimize financial distress and enhance food availability by reducing intermediaries in the supply chain. Key stakeholders often face barriers to transparency, security, and inefficiencies in their agricultural practices (Srikanth et al., 2023). The adoption of blockchain can address these challenges by reducing the risk of agronomic data being tampered with, while supporting data integrity (Xiong et al., 2020). From an environmental sustainability perspective, data from IoT sensors can help mitigate the overexploitation of natural resources. As the blockchain network preserves the integrity of the data recorded by IoT-based sensors, it leverages a decentralized authentication mechanism (Mohanta et al., 2021). Also, through blockchain, a natural disaster response system can be potentially managed, as it can help in supporting auditability, traceability, and coordination (Suma et al., 2017), thereby reducing losses and improving predictability. From an economic perspective, BDSA can promote global trade by encouraging e-commerce for agricultural products (Torky & Hassanein, 2020). It can significantly amplify the profit margin within the BDSA for certain relevant stakeholders by reducing intermediaries in the supply chain from the field to the food (Sajja et al., 2023). In blockchain, data are verified via a consensus mechanism (Demestichas et al., 2020), which might consequently reduce financial disputes in BDSA systems. Given that blockchain is a decentralized digital ledger system (Yli-Huumo et al., 2016), it can enable BDSA stakeholders to buy and exchange goods, thereby further supporting greater financial independence.

Using the TBL framework, we have also identified key challenges and limitations that can hinder the adoption of BDSA across social, environmental, and economic dimensions. The adoption of BDSA can face a multitude of social challenges, which can arise from a lack of knowledge, accessibility to digital infrastructure, and regulatory issues (Nagariya et al., 2022; Potts, 2019). The absence of active participation and trust among associated stakeholders can hinder the potential utilization of blockchain in BDSA (Xiong et al., 2020). Also, numerous environmental barriers can pose significant hurdles. The lack of proper internet connectivity, particularly in remote areas, can slow down the adoption of blockchain solutions for smart agriculture (Pakeerathan, 2023b). Also, the presence of untrustworthy data providers in the supply chain can reduce sustainable agriculture practices (Sadiku et al., 2021). Apart from these, certain economic challenges can hinder the adoption of BDSA, such as the requirement of high energy costs (Potts, 2019), especially in proof-of-work, and hardware incompatibility, among others (Xiong et al., 2020).

This chapter explores how blockchain technology can sustainably transform conventional farming practices through the TBL framework. The remainder of the chapter is structured as follows: Section 2 discusses the attributes of smart agriculture. Section 3 addresses the challenges smart agriculture faces in cybersecurity, connectivity, and the availability of skilled workers. Section 4 discusses blockchain, including its history, evolution, and applications in various sectors. Section 5 presents an in-depth analysis of the impact of blockchain technology on smart agriculture, utilizing the TBL framework to examine its social, environmental, and economic implications. Section 6 identifies key challenges faced by BDSA using the TBL framework, including untrustworthy data providers, regulatory barriers, high energy costs, and system incompatibility. Moving forward, Section 7 discusses additional challenges in conceptualizing BDSA, including practical limitations and implementation issues. Section 8 highlights key outcomes and takeaways, summarizes the critical findings, and offers insights into the future development of BDSA.

2. Smart Agriculture System

The United Nations has estimated that the global population could reach 9.7 billion by 2050, driving a significant demand for agricultural products (Pakeerathan, 2023b). To meet the increasing demand for farm produce, relevant practitioners are adopting innovative methods, aided by emerging technologies, to enhance the productivity of farm products. One such innovative method is smart agriculture; it refers to a farming paradigm that leverages emerging technologies to improve the quality and quantity of agricultural produce (Goel et al., 2021). Smart farming enables farmers to optimize resource use, including fertilizers, pesticides, and human resources. Smart farming encompasses various data transformation methodologies, including collection, management, automation, transmission, and decision-making (Issad et al., 2019). Agricultural data, including soil and meteorological conditions, are collected using various equipment, such as drones and sensors. Farmers use these data for planning soil and water use, monitoring crop health, limiting the use of pollutants (e.g., pesticides and herbicides), and improving the quality of production (Issad et al., 2019). The installation of a wireless sensor network (WSN) in agriculture has influenced farming practices, increasing farmers' effectiveness and efficiency. A WSN helps track variables in farming fields, such as soil conditions, atmospheric conditions, and biomass. It can also help to monitor temperature and humidity variables (Muangprathub et al., 2019). The sensors' data are deployed to the embedded system, where it is analyzed, and the decisions are directly transferred to the equipment for action (Pakeerathan, 2023a). Integrating these components can potentially transform traditional farming practices into data-driven, efficient, and sustainable operations. Implementing smart agriculture can benefit the stakeholders involved. They can efficiently manage the quantity of pesticides and herbicides used in their farmlands. Trust among consumers is also enhanced by the traceability and efficiency offered by BDSA from seed to food.

Smart agriculture enhances traceability, trust, and optimized use of pollutants and water, while also improving the quality of production. Some of the prominent advantages of smart agriculture are summarized in Fig. 12.1 (Hasan et al., 2024; Tzachor et al., 2022; Walter et al., 2017).

3. Issues and Challenges for the Smart Agriculture System

Food is fundamental to the survival of all living beings, and the agriculture and food sector is a significant source of global employment. A strong agriculture sector symbolizes a socially and economically stable nation. Farmers used sticks and sickles to cultivate crops and hunt animals in the early years. Today, farmers are adopting innovative technologies to enhance productivity. With these advanced technologies, farmers can remotely monitor their farms using control devices. Nowadays, agricultural practitioners are gradually incorporating genetically enhanced seeds to mitigate diseases. These genetically modified seeds enhance crop quality, resulting in higher yields and greater production. These improved qualities of crops and harvests can reduce global food scarcity (Idoje et al., 2021).

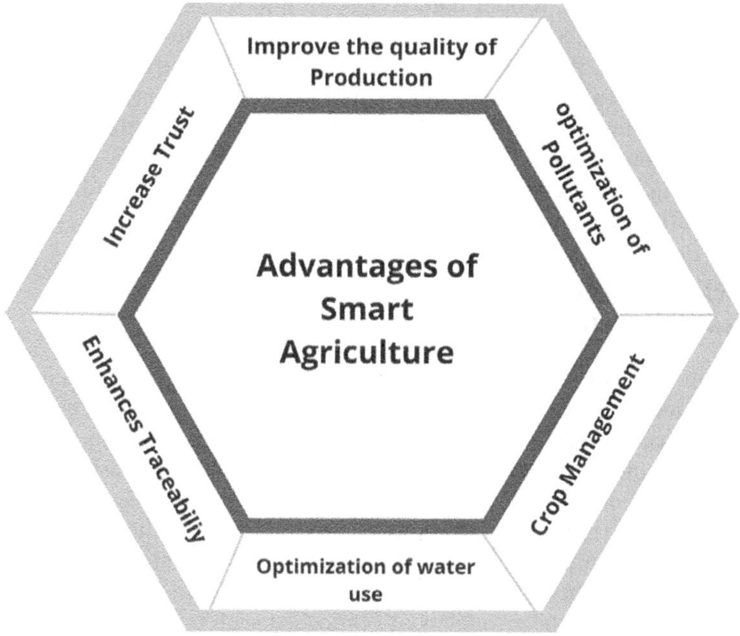

Fig. 12.1. Advantages of Smart Agriculture.

Adopting pioneering technologies in farming practices enhances crop production and quality while minimizing environmental impact and production costs.

The adoption of smart agriculture has the potential to deliver significant benefits by enhancing crop production while minimizing waste. However, there are specific challenges to smart agriculture. Smart agricultural systems are distributed and prone to cyberattacks, including data-integrity attacks, denial-of-service attacks, and other disruptions that could compromise the system's privacy, integrity, and availability (Singh et al., 2022). Smart farming relies on various IoT-based devices, each utilizing different protocols or standards to perform specific tasks. As a result, compatibility issues can occur when connecting devices running different firmware and operating systems. Such an issue can also arise when deploying software to hardware, as some controller boards are unable to handle artificial intelligence models (Pakeerathan, 2023b). Connecting many devices to the internet can be a significant problem. The future of agriculture may increasingly incorporate decentralized architectures (see Fig. 12.2) (Shukla et al., 2020), and the success of IoT-based devices depends on proper internet connectivity. The operation of smart devices requires experienced and highly skilled workers. However, the need for additional skills in handling these smart devices poses challenges to the adoption of smart agriculture (Pakeerathan, 2023b; Singh et al., 2022). The lack of standardization of farm devices is also a significant hurdle to smart agriculture. For the effectiveness of smart farming, data collection and security are crucial for improving productivity. However, proper data collection and security have become critical in smart farming. Smart agriculture enables farmers to monitor their farms remotely. However, the researcher must focus on effectively tracking the animals' locations, health, and changes in their behavior patterns within the farm to collect real-time animal data. Different plants require different soil types for optimal growth, yet accurately determining soil conditions is challenging in smart farming. Research must be conducted to determine soil conditions before crop cultivation to ensure a productive harvest (Idoje et al., 2021).

4. Blockchain Technology: Concept and Real-world Applications

Blockchain is one of the most widely discussed emerging technologies in the current era of technological advancement (Singh et al., 2021). Its inherent attribute concerning decentralization and peer-to-peer technology can aid in tracking and verification of digital transactions (Imteaj et al., 2021; Jena & Dash, 2021).

Decentralization is a critical aspect of blockchain, as it implies the delegation, dispersion, or distribution of power away from a central authority (Hoffman et al., 2020; Shukla et al., 2020). Refer to Fig. 12.2 for a schematic representation of the centralized and decentralized systems.

All contributors in the arrangement acknowledge all transactions executed on the blockchain. Each transaction entered into a blockchain cannot be erased, as it is an immutable ledger (Sheth & Dattani, 2019). Each data record is replicated across multiple network nodes, which minimizes the chances of loss and thereby enhances its trustworthiness. The development of blockchain is attributed to several historical events that have shaped the current understanding. The history of blockchain development is summarized in Table 12.1.

Some of the blockchain's prominent features, which can encourage BDSA, are shown in Table 12.2.

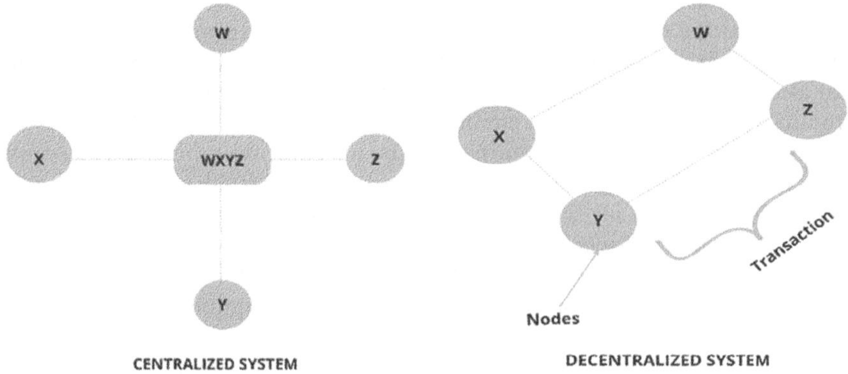

Fig. 12.2. The Schematic Diagram for Centralized and Decentralized Systems.

Table 12.1 Overview of Blockchain History, the Timeline Provided Is Based on Guo and Yu (2022).

Year	Event Description	Reference
1982	David Chaum proposed a blockchain-like protocol in his PhD thesis titled "Computer Systems Established, Maintained, and Trusted by Mutually Suspicious Groups."	Panda et al. (2020)
1998	Wei Dai published a proposal for "b-money," introducing the concept of decentralized digital currency. It included ideas such as using cryptography for secure transactions and maintaining a distributed ledger.	Raskin and Yermack (2018)
2008	Bitcoin, a decentralized digital currency, was introduced under the pseudonym Satoshi Nakamoto, a possibly fictitious identity.	Crosby (2016); Sarmah (2018)
2013	Ethereum, an open-source blockchain platform, was proposed. It enables developers to build and deploy smart contracts and decentralized applications.	Popovski et al. (2018)
2015	The term "blockchain" gained widespread attention due to its association with Bitcoin and other cryptocurrencies.	Buitenhek (2016)
2020–2022	Ethereum 2.0 introduced significant upgrades to the Ethereum network, enhancing its speed, scalability, and security.	Allam (2025)

A survey by PwC shows that 46% of the financial service industry has adopted blockchain technology, followed by industrial products and manufacturing (12%), energy and utilities (12%), and healthcare (11%), the bottom three are government (8%), retail and consumer (4%), and entertainment and media (1%) (Taskinsoy, 2019). Blockchain with smart contracts is crucial for the financial industry, as it enables fast settlement and is immune to data tampering. In the healthcare sector, patient records contain sensitive information intended to be kept private and secure for the patient's proper diagnosis and well-being (Shukla et al., 2020). To keep patients' data secure, blockchain technology is of great interest to the healthcare industry. Blockchain can significantly help by tracing

Table 12.2 Prominent blockchain features for BDSA implementation, as elucidated by Imteaj et al. (2021). In this table, "BKC" denotes "blockchain key characteristic" and the numbers identify the individual characteristics.

Abbreviation	Feature	Description	Reference(s)
BKC-1	Decentralization	Blockchain technology can enhance security in BDSA by leveraging decentralization, which distributes control across multiple nodes and reduces reliance on a single authority.	Xiong et al. (2020)
BKC-2	Transparency	Transactions on the blockchain are validated through a consensus mechanism, ensuring transparency across the process.	Demestichas et al. (2020)
BKC-3	Traceability	Traceability enhances trust and confidence among retailers and customers regarding the products they purchase.	Mitani and Otsuka (2020)
BKC-4	Immutability	Because blockchain is immutable, it can provide regulatory agencies with reliable, accurate information for efficient regulation.	Xiong et al. (2020)
BKC-5	Safety	Using blockchain, industry officials can quickly identify the source of the contaminant and determine the scope of the affected products, thereby enhancing product safety.	Aldan (2019)
BKC-6	Pseudonymous	On blockchain platforms, user accounts can be created without any identity or authorization process. This allows users to use a fictitious name.	Berryhill et al. (2018)
BKC-7	Persistency	Persistence in blockchain helps ensure that recorded data remain permanent and unalterable.	Vaigandla et al. (2023)
BKC-8	Auditability	Auditability can enable a transparent and immutable record of all transactions.	Farooq et al. (2020)

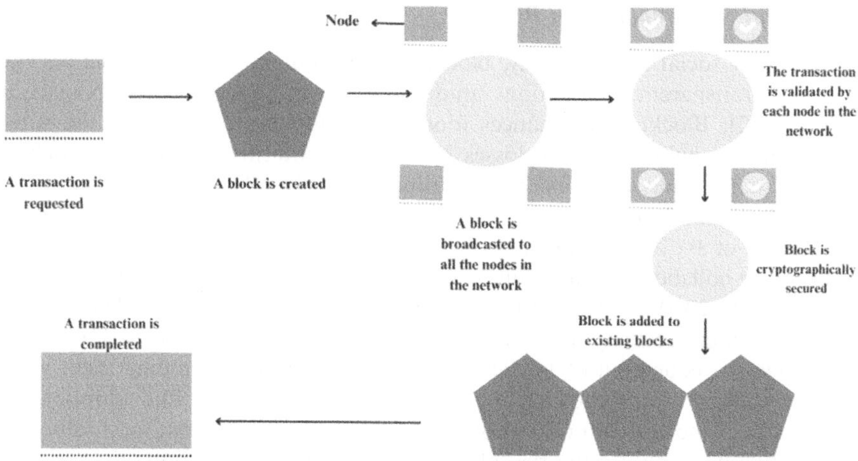

Fig. 12.3. Blockchain's Transaction Mechanism.

where food comes from to ensure safety, managing information systems, supporting trade and finance, and providing crop certification and insurance (Yadav & Singh, 2019). It can also enhance the integrity of recorded agronomic data; though its reliability still depends on the trustworthiness of the data sources.

Food epidemics in the last 20 years, such as the 2006 Escherichia coli outbreak in the United States and the Sanlu milk scandal in China, enhanced consumers' concerns regarding food provenance and quality (Demestichas et al., 2020). Despite the advanced technologies already in use, the majority of systems remain centralized and outdated. In this regard, distributed ledger technologies such as blockchain are a reliable solution to this issue. Upon consulting the existing literature, we have summarized the core mechanism through which transactions are processed within a blockchain; refer to Fig. 12.3 (Sheth & Dattani, 2019; Tanwar, 2022; Yaga et al., 2018).

5. TBL-based Examination of BDSA

Agriculture lies at the heart of human progress, and a country's economic growth depends on the strength of its agricultural sector. Today, technology has impacted numerous sectors of industry. In the modern world, many aspects are automated, including robotics and the stock market. The global farming industry continues to grow each year to meet the increasing demand for food for the world's population. Prior research indicates a growing adoption of innovative farming methods aimed at enhancing productivity, increasing yield, and reducing waste and environmental impact (Biswas et al., 2021).

In this section, we aim to study the impact of BDSA through the TBL framework. The TBL is a sustainability evaluation framework that incorporates aspects of a specific ecosystem into social, environmental, and economic dimensions (Slaper & Hall, 2011). TBL was introduced in 1994 and has been applied in multiple domains to assess key aspects of sustainability (Loviscek, 2020).

5.1. Social Sustainability in BDSA

In line with social sustainability, blockchain promotes fair trade practices by ensuring transparent transactions among agricultural practitioners (Nagariya et al., 2022). Blockchain enhances food availability by rationalizing the supply chain and helps minimize losses by reducing intermediaries (Demestichas et al., 2020). There has been concern in traditional agricultural spaces for a transparent and secure method to trace the provenance and authenticity of farm goods (Sandeep Kumar et al., 2021). BDSA systems can address this concern by facilitating stakeholder collaboration and minimizing disputes through efficient traceability (Xiong et al., 2020). Furthermore, integrating IoT devices with blockchain can enhance farmers' security (Lin et al., 2018). From an educational lens, BDSA can encourage the education of participating stakeholders by equipping them with the current trends and knowledge of smart farming practices. The adoption of blockchain in the agricultural industry can reduce concerns about food safety in the supply chain, reduce the risk of fraud and data counterfeiting, and thereby strengthen security (Lin et al., 2018). Thus, BDSA systems can enhance employability by creating novel roles in farming and data warehousing industries. BDSA setups can facilitate the digital registration of agricultural practitioners' land and assets. As a consequence, it could help them to establish an online presence and protect them from legal and financial risks.

5.2. Environmental Sustainability in BDSA

The environmental aspects of BDSA are of significant importance. For instance, IoT-based sensors (such as soil and moisture, Passive Infrared (PIR), and temperature) (Suma et al., 2017) are used to capture agronomic data and climate conditions. These data are sent to connected devices for processing and decision-making. The blockchain network is used to preserve the integrity and traceability of the data using a decentralized authentication protocol, and final decisions are recorded as transactions in the blockchain (Mohanta et al., 2021). The data recorded by IoT sensors can support more balanced natural resource usage and may help reduce the overexploitation of soil and water (Biswas et al., 2021). It can support more responsible use of inputs by enabling traceability of produce from the field to the food. Blockchain and IoT-powered innovative greenhouse systems also help reduce energy consumption through careful resource management (Lin et al., 2018). Environmental monitoring in BDSA via blockchain can help farmers to quickly adapt and consequently take action in response to natural disasters like droughts and floods, ensuring the minimization of possible damage (Suma et al., 2017). Furthermore, blockchain enhances sustainable management throughout a product's lifecycle by recording every step in the value chain (Xiong et al., 2020).

5.3. Economic Sustainability in BDSA

From an economic sustainability perspective, blockchain can increase profit margin for agriculture practitioners by reducing the number of middlemen in the supply chain from fields to consumers (Sajja et al., 2023). The transparency provided in the supply chain through blockchain enhances the market value of agricultural produce (Lin et al., 2018). Agricultural subsidies can be accurately tracked and distributed to the farmers within BDSA (Torky & Hassanein, 2020). A blockchain is an immutable ledger that operates through a consensus mechanism. This allows stakeholders to trust the data seen or added to the blockchain. Transactions in the blockchain can be verified by the network of participants through a consensus method, thereby reducing financial disputes (Demestichas et al., 2020). Agricultural practitioners can benefit from easier access to agricultural loans, as their records are easily verifiable through blockchain technology. Blockchain facilitates global trade by enhancing the e-commerce of farming commodities (Sajja et al., 2023; Torky & Hassanein, 2020). Apart from this, multi-echelon production allocation challenges can be effectively addressed by blockchain, as it can support the implementation of carbon taxation policies, which may help in economic incentivization for sustainable practices (Nagariya et al., 2022).

Blockchain technology is evolving and could create a profound impact on the economy, society, and the environment. BDSA encourages farming practices that are more efficient, reliable, and transparent. Fig. 12.4 summarizes potential applications of BDSA (Lin et al., 2018; Nagariya et al., 2022; Sajja et al., 2023). However, BDSA faces several limitations, including limited awareness among key stakeholders, limited scalability, high energy requirements, limited access to blockchain-compatible infrastructure, and uncertain regulatory frameworks.

Fig. 12.4. Potential Applications of Blockchain in Smart Agriculture.

6. Analyzing Challenges of BDSA Using the TBL Framework

Based on several research articles examining the impact of blockchain on smart agriculture, we identified several concerning challenges with its applicability. These challenges can be elucidated through the TBL framework's pillars.

6.1. Social Challenges in BDSA

From a social perspective, the large-scale adoption of BDSA systems can become problematic due to a lack of awareness among stakeholders. As blockchain is an emerging technology, a considerable gap exists in the literature, particularly for stakeholders in agro-based industries. This dearth in knowledge elicitation and training limits stakeholders from acknowledging the potential of blockchain, which may further delay its adoption (Potts, 2019; Ramdinthara, 2020). Also, it requires active participation from most of the stakeholders involved (farmers, policymakers, and consumers) to realize the potential impact of BDSA fully (Sadiku et al., 2021; Xiong et al., 2020). Moreover, regulatory issues pose significant challenges to the adoption of blockchain technology. To streamline its adoption, proper government guidelines are essential, enabling industries to implement blockchain in the agro-based sector effectively (Nagariya et al., 2022).

6.2. Environmental Challenges in BDSA

As smart agriculture demands seamless internet connectivity and an interconnected IoT ecosystem of devices, sensors, and drones, this poses a challenge, especially in rural areas where internet connectivity remains limited (Pakeerathan, 2023b; Prasad et al., 2022). This can hamper BDSA's scalability in practice, even though blockchain is often praised for its data security and transparency. However, its effectiveness is still doubted when there is one or two untrustworthy data providers in the supply chain, which may introduce inaccuracies into the data (Sadiku et al., 2021; Sandeep Kumar et al., 2021). This reduces the environmental benefits of blockchain in monitoring and sustainable agricultural practices. Blockchain operations often require significant computational power; thus, large-scale adoption could lead to high energy consumption, potentially further hindering the achievement of the sustainable development goals.

6.3. Economic Challenges in BDSA

From an economic perspective, adopting blockchain incurs significant energy costs, as it requires computational resources to process concurrent transactions and their respective verifications. On the other hand, profit margins in the agricultural domain are considerably low (Potts, 2019; Ramdinthara, 2020), which can be further affected by energy costs. This can create dissonance among the public regarding the implementation of large-scale BDSA systems. Additionally, conventional agricultural setups are heavily reliant on legacy infrastructure.

These systems can pose computational incompatibility with the technical stack of emerging technologies, such as blockchain. While integrating these technologies with these systems requires commendable concentrated effort in terms of finances and time, this effort leads to an imperative increase in the adoption cost of blockchain. This functions as a significant barrier for agricultural corporations, particularly those governed in developing regions, where the availability of resources is often minimal (Sadiku et al., 2021; Xiong et al., 2020). Also, the supply chain management in the agricultural sector is complex due to multifaceted challenges in agricultural production across various fronts, including weather conditions and pest control. Thus, integrating real-time data about pest control, weather patterns, and diseases into intelligent blockchain systems can be economically infeasible (Sadiku et al., 2021). Thus, BDSA systems can be prone to financial risk given the current status of blockchain.

Although the positive implications of blockchain technology have attracted significant attention from research and practice-based communities, highlighting its shortcomings may further enhance its practical applications, such as BDSA.

7. Conclusion and Future Directions

BDSA has the potential to significantly influence the agriculture domain by addressing the social, environmental, and economic dimensions, as identified through the TBL framework in this study. As blockchain fosters security, transparency, and trust among stakeholders, it may help BDSA enhance food availability and support sustainable farming practices. Additionally, it can improve market access for farmers by enabling transparent, secure transactions. BDSA can strengthen trust among stakeholders, which may help relevant practitioners educate them about smart farming solutions. Additionally, the overexploitation of natural resources can be minimized through BDSA, as its inherent transparency can effectively promote efficient resource use.

However, this chapter also identifies and summarizes several limitations of BDSA, as derived from the existing literature, using the TBL framework. From a social perspective, the lack of trust and awareness regarding blockchain systems in BDSA can hinder its scalability and sustainable implementation. Data security is of paramount concern, as weak implementation of technical infrastructure might lead to data leakage to unauthorized parties. From an environmental perspective, blockchain's high energy consumption poses significant sustainability challenges. Moreover, the use of inaccurate data within blockchain systems can impair decision-making related to environmental sustainability. Economically, the initial implementation of blockchain depends on substantial investments in infrastructure, technology, and training. It also poses challenges for those reliant on traditional agricultural markets, notably distributors and brokers. The adoption of novel technical premises within conventional spaces without proper training and education can also lead to disorientation among the masses.

To address these challenges, research and technological advancements are crucial for making blockchain systems more sustainable, efficient, scalable, and cost-effective. Researchers must accelerate progress from data collection to implementation by addressing blockchain's inherent flaws through continuous research. Effective communication among stakeholders, including farmers, policymakers, and consumers, is crucial, given the diverse range of blockchain systems in use. Establishing standard protocols for smart contracts and consensus mechanisms can facilitate more optimal and secure data exchange, thereby fostering collaboration among agricultural stakeholders. Additionally, collective focus should be channeled on developing optimized computational protocols that minimize power consumption and reduce environmental impact, thereby further strengthening the applicability of BDSA systems in practical bandwidth applications. Therefore, the key takeaways of this study can be elucidated as follows:

A. BDSA can enhance aspects of sustainability across social, environmental, and economic dimensions by improving resource management, reducing intermediaries, and fostering transparency.
B. BDSA-enabled systems can help reduce the overexploitation of resources, support disaster resilience, and improve food availability by ensuring traceability and trust among the stakeholders concerned.
C. While BDSA reduces costs and boosts profitability, high energy requirements, infrastructure incompatibility, and regulatory gaps hinder large-scale adoption.
D. Research and standardized protocols are vital when scalability is concerned. There is also a critical need for energy-efficient protocols and for resolving regulatory barriers to ensure the effective and widespread implementation of BDSA ecosystems.

This chapter presents a comprehensive, holistic understanding of the intersection of blockchain and smart agriculture, highlighting its benefits and key challenges. We believe this study may be helpful to researchers, practitioners, and individuals with a vested interest in understanding this intersection of smart agriculture and blockchain, thereby enhancing a nuanced understanding.

References

Aldag, M. C. (2019). The use of blockchain technology in agriculture. *Zeszyty Naukowe Uniwersytetu Ekonomicznego w Krakowie*, *982*(4), 7–17.

Allam, S. (2025). Financial institutions' integration of blockchain technology for cross-border payment optimization: A systematic analysis. *Journal of Computer Science and Technology Studies*, *7*(8), 70–76. https://doi.org/10.32996/jcsts.2025.7.8.9

Berryhill, J., Bourgery, T., & Hanson, A. (2018). *Blockchains unchained: Blockchain technology and its use in the public sector*. https://www.oecd-ilibrary.org/governance/blockchains-unchained_3c32c429-en

Biswas, M., Akhund, T. M. N. U., Ferdous, M. J., Kar, S., Anis, A., & Shanto, S. A. (2021). Biot: Blockchain based smart agriculture with internet of thing. In *2021 Fifth world conference on smart trends in systems security and sustainability (WorldS4)* (pp. 75–80). IEEE. https://ieeexplore.ieee.org/abstract/document/9513998/

Buitenhek, M. (2016). Understanding and applying Blockchain technology in banking: Evolution or revolution? *Journal of Digital Banking*, *1*(2), 111–119.

Crosby, M. (2016). BlockChain technology: Beyond bitcoin. *Applied Innovation Review, 2*, 1–19.

Demestichas, K., Peppes, N., Alexakis, T., & Adamopoulou, E. (2020). Blockchain in agriculture traceability systems: A review. *Applied Sciences, 10*(12), 4113. https://doi.org/10.3390/app10124113

Farooq, M. S., Khan, M., & Abid, A. (2020). A framework to make charity collection transparent and auditable using blockchain technology. *Computers & Electrical Engineering, 83*, 106588.

Goel, R. K., Yadav, C. S., Vishnoi, S., & Rastogi, R. (2021). Smart agriculture – Urgent need of the day in developing countries. *Sustainable Computing: Informatics and Systems, 30*, 100512.

Guo, H., & Yu, X. (2022). A survey on blockchain technology and its security. *Blockchain: Research and Applications, 3*(2), 100067.

Hasan, H. R., Musamih, A., Salah, K., Jayaraman, R., Omar, M., Arshad, J., & Boscovic, D. (2024). Smart agriculture assurance: IoT and blockchain for trusted sustainable produce. *Computers and Electronics in Agriculture, 224*, 109184.

Hoffman, M. R., Ibáñez, L.-D., & Simperl, E. (2020). Toward a formal scholarly understanding of blockchain-mediated decentralization: A systematic review and a framework. *Frontiers in Blockchain, 3*. https://doi.org/10.3389/fbloc.2020.00035

Idoje, G., Dagiuklas, T., & Iqbal, M. (2021). Survey for smart farming technologies: Challenges and issues. *Computers & Electrical Engineering, 92*, 107104.

Imteaj, A., Hadi Amini, M., & Pardalos, P. M. (2021). Introduction to blockchain technology. In A. Imteaj, M. H. Amini, & P. M. Pardalos (Eds.), *Foundations of blockchain* (pp. 3–13). Springer International Publishing. https://doi.org/10.1007/978-3-030-75025-1_1

Issad, H. A., Aoudjit, R., & Rodrigues, J. J. (2019). A comprehensive review of data mining techniques in smart agriculture. *Engineering in Agriculture, Environment and Food, 12*(4), 511–525.

Jena, A. K., & Dash, S. P. (2021). Blockchain technology: Introduction, applications, challenges. In S. K. Panda, A. K. Jena, S. K. Swain, & S. C. Satapathy (Eds.), *Blockchain technology: Applications and challenges* (Intelligent Systems Reference Library, Vol. 203). Springer. https://doi.org/10.1007/978-3-030-69395-4_1

Jena, A. K., & Dash, S. P. (2021). Blockchain technology: Introduction, applications, challenges. In *Blockchain technology: Applications and challenges* (pp. 1–11). Springer. https://doi.org/10.1007/978-3-030-69395-4_1

Laurence, T. (2019). *Introduction to blockchain technology*. Van Haren. https://books.google.com/books?hl=en&lr=&id=uD-4DwAAQBAJ&oi=fnd&pg=PR4&dq=Blockchain+Introduction&ots=SWSYHNTGeE&sig=BvCitUhggcyQDxzrsw7UokjqFIE

Lin, J., Shen, Z., Zhang, A., & Chai, Y. (2018). Blockchain- and IoT-based food traceability for smart agriculture. In *Proceedings of the 3rd international conference on crowd science and engineering* (pp. 1–6). ACM. https://doi.org/10.1145/3265689.3265692

Loviscek, V. (2020). Triple bottom line toward a holistic framework for sustainability: A systematic review. *Revista de Administração Contemporânea, 25*, e200017. https://doi.org/10.1590/1982-7849rac2021200017.en

Mitani, T., & Otsuka, A. (2020). Traceability in permissioned blockchain. *IEEE Access, 8*, 21573–21588.

Mohanta, B. K., Chedup, S., & Dehury, M. K. (2021). Secure trust model based on blockchain for internet of things enable smart agriculture. In *2021 19th OITS international conference on information technology (OCIT)* (pp. 410–415). IEEE. https://ieeexplore.ieee.org/abstract/document/9719520/

Muangprathub, J., Boonnam, N., Kajornkasirat, S., Lekbangpong, N., Wanichsombat, A., & Nillaor, P. (2019). IoT and agriculture data analysis for smart farm. *Computers and Electronics in Agriculture, 156*, 467–474.

Nagariya, R., Mukherjee, S., Baral, M. M., Patel, B. S., & Venkataiah, C. (2022). The challenges of blockchain technology adoption in the agro-based industries. *International Journal of Mathematical, Engineering and Management Sciences*, *7*(6), 949.

Pakeerathan, K. (Ed.). (2023a). *Smart agriculture for developing nations: Status, perspectives and challenges*. Springer Nature Singapore. https://doi.org/10.1007/978-981-19-8738-0

Pakeerathan, K. (2023b). Smart agriculture: Special challenges and strategies for island states. In K. Pakeerathan (Ed.), *Smart agriculture for developing nations* (pp. 251–258). Springer Nature Singapore. https://doi.org/10.1007/978-981-19-8738-0_18

Panda, S. K., Elngar, A. A., Balas, V. E., & Kayed, M. (2020). *Bitcoin and blockchain: History and current applications*. CRC Press. https://books.google.com/books?hl=en&lr=&id=iYr1DwAAQBAJ&oi=fnd&pg=PT14&dq=history+of+blockchain+technology&ots=JqbCsD-11A&sig=V6pWSv66qhjT7n_8fF1SH_7UXV4

Popovski, L., Soussou, G., & Webb, P. B. (2018). A brief history of blockchain. *Legaltech News*, *2018*, 3.

Potts, J. (2019). *Blockchain in agriculture*. SSRN 3397786. https://papers.ssrn.com/sol3/papers.cfm?abstract_id=3397786

Pranto, T. H., Noman, A. A., Mahmud, A., & Haque, A. B. (2021). Blockchain and smart contract for IoT enabled smart agriculture. *PeerJ Computer Science*, *7*, e407.

Prasad, C. G., Mallareddy, A., Pounambal, M., & Velayutham, V. (2022). Edge computing and blockchain in smart agriculture systems. *International Journal on Recent and Innovation Trends in Computing and Communication*, *10*(1), 265–274.

Ramdinthara, I. Z. & Shanthi Bala P. (2020). Issues and Challenges in Smart Farming for Sustainable Agriculture. In N. Pradeep, S. Kautish, C. Nirmala, V. Goyal, & S. Abdellatif (Eds.), *Modern techniques for agricultural disease management and crop yield prediction* (pp. 1–22). IGI Global. https://doi.org/10.4018/978-1-5225-9632-5.ch001

Raskin, M., & Yermack, D. (2018). Digital currencies, decentralized ledgers and the future of central banking. In P. Conti-Brown & R. M. Lastra (Eds.), *Research handbook on central banking* (Research Handbooks in Financial Law series, pp. 474–486). Edward Elgar Publishing. https://doi.org/10.4337/9781784719227.00028

Sadiku, M. N., Eze, K. G., & Musa, S. M. (2021). *Blockchain in agriculture*. https://procon.bg/bg/system/files/4904_blockchain_in_agriculture_preview.pdf

Sajja, G. S., Rane, K. P., Phasinam, K., Kassanuk, T., Okoronkwo, E., & Prabhu, P. (2023). Towards applicability of blockchain in agriculture sector. *Materials Today: Proceedings*, *80*, 3705–3708. https://doi.org/10.1016/j.matpr.2021.07.366

Sandeep Kumar, M., Maheshwari, V., Prabhu, J., Prasanna, M., & Jothikumar, R. (2021). Applying blockchain in agriculture: A study on blockchain technology, benefits, and challenges. In A. Suresh & S. Paiva (Eds.), *Deep learning and edge computing solutions for high performance computing* (pp. 167–181). Springer International Publishing. https://doi.org/10.1007/978-3-030-60265-9_11

Sarmah, S. S. (2018). Understanding blockchain technology. *Computer Science and Engineering*, *8*(2), 23–29.

Sheth, H., & Dattani, J. (2019). Overview of blockchain technology. *Asian Journal for Convergence in Technology (AJCT)*. ISSN-2350-1146. http://asianssr.org/index.php/ajct/article/view/728

Shukla, R. G., Agarwal, A., & Shukla, S. (2020). Chapter 10 – Blockchain-powered smart healthcare system. In S. Krishnan, V. E. Balas, E. G. Julie, Y. H. Robinson, S. Balaji, & R. Kumar (Eds.), *Handbook of research on blockchain technology* (pp. 245–270). Academic Press. https://doi.org/10.1016/B978-0-12-819816-2.00010-1

Singh, G., Garg, V., & Tiwari, P. (2021). Introduction to blockchain technology. In R. Agrawal & N. Gupta (Eds.), *Transforming cybersecurity solutions using blockchain* (pp. 1–18). Springer. https://doi.org/10.1007/978-981-33-6858-3_1

Singh, G., Kalra, N., Yadav, N., Sharma, A., & Saini, M. (2022). Smart agriculture: A review. *Siberian Journal of Life Sciences and Agriculture*, *14*(6), 423–454.

Slaper, T. F., & Hall, T. J. (2011). The triple bottom line: What is it and how does it work. *Indiana Business Review*, *86*(1), 4–8.

Srikanth, M., Mohan, R. J., & Naik, M. C. (2023). Blockchain-based consensus for a secure smart agriculture supply chain. *European Chemical Bulletin*, *12*(4), 8669–8678.

Suma, N., Samson, S. R., Saranya, S., Shanmugapriya, G., & Subhashri, R. (2017). IoT based smart agriculture monitoring system. *International Journal on Recent and Innovation Trends in Computing and Communication*, *5*(2), 177–181.

Tanwar, S. (2022). Introduction to blockchain technology. In S. Tanwar (Ed.), *Blockchain technology: From theory to practice* (pp. 1–41). Springer Nature. https://doi.org/10.1007/978-981-19-1488-1_1

Taskinsoy, J. (2019, October 8). *Blockchain: A misunderstood digital revolution. Things you need to know about blockchain* [SSRN scholarly paper]. https://doi.org/10.2139/ssrn.3466480

Torky, M., & Hassanein, A. E. (2020). Integrating blockchain and the internet of things in precision agriculture: Analysis, opportunities, and challenges. *Computers and Electronics in Agriculture*, *178*, 105476.

Tzachor, A., Devare, M., King, B., Avin, S., & ÓhÉigeartaigh, S. (2022). Responsible artificial intelligence in agriculture requires systemic understanding of risks and externalities. *Nature Machine Intelligence*, *4*(2), 104–109.

Vaigandla, K. K., Karne, R., Siluveru, M., & Kesoju, M. (2023). Review on blockchain technology: Architecture, characteristics, benefits, algorithms, challenges and applications. *Mesopotamian Journal of CyberSecurity*, *2023*, 73–84.

Walter, A., Finger, R., Huber, R., & Buchmann, N. (2017). Smart farming is key to developing sustainable agriculture. *Proceedings of the National Academy of Sciences*, *114*(24), 6148–6150. https://doi.org/10.1073/pnas.1707462114

Xiong, H., Dalhaus, T., Wang, P., & Huang, J. (2020). Blockchain technology for agriculture: Applications and rationale. *Frontiers in Blockchain*, *3*. https://doi.org/10.3389/fbloc.2020.00007

Yadav, V. S., & Singh, A. R. (2019). A systematic literature review of blockchain technology in agriculture. In *Proceedings of the international conference on industrial engineering and operations management* (Vol. 23, pp. 973–981). IEOM Society International.

Yaga, D., Mell, P., Roby, N., & Scarfone, K. (2018). *Blockchain technology overview* (NIST IR 8202). https://doi.org/10.6028/NIST.IR.8202

Yli-Huumo, J., Ko, D., Choi, S., Park, S., & Smolander, K. (2016). Where is current research on blockchain technology? – A systematic review. *PLOS ONE*, *11*(10), e0163477. https://doi.org/10.1371/journal.pone.0163477

Chapter 13

Unlocking the Potential of Millets in India: A Climate-smart Approach to Sustainable Livelihood and Global Trade

Isha Jaswal[a], *Shivani Mehta*[b] *and Palak Rawat*[c]

[a] *Vivekananda School of Economics, Vivekananda Institute of Professional Studies – Technical Campus, New Delhi, India*
[b] *Department of Economics, Jagan Institute of Management Studies, Rohini, Delhi, India*
[c] *Department of Economics, Manav Rachna International Institute of Research and Studies, Haryana, India*

Abstract

Millets, often called "climate-smart grains," have emerged as pivotal in addressing the intertwined challenges of food security, climate resilience, and sustainable livelihoods. In line with the sustainable development goals (SDGs), including SDG 2 (Zero Hunger), SDG 3 (Good Health and Well-being), SDG 13 (Climate Action), and SDG 8 (Decent Work and Economic Growth), this chapter examines the diverse potential of millets in India. Millets are essential to climate-smart agriculture (CSA) because of their low water requirements and capacity to flourish on nutrient-poor, arid soils (Sikiru et al., 2023). Through their production, processing, and sale, these grains guarantee steady yields under challenging weather conditions and help rural households by generating long-term employment prospects. The study focuses on women-led initiatives across Indian states that increase revenue and advance gender parity, highlighting the transformative power of value addition in millet-based businesses. Additionally, the report describes how new millet products and enhanced market connections support rural economies and meet the growing demand for health-conscious diets around the world. Lastly, this chapter examines the potential of the millet industry, emphasizing India's potential to position

Sustainable and Smart Farming Eco-systems:
Changing Landscapes in the Age of Technology, 183–198

doi:10.1108/978-1-83608-092-320261013

itself as a leader in the global millet market, through general equilibrium analysis, studying the associated spillover effects and complexities. By integrating millets into CSA frameworks, fostering value chain development, and scaling their trade potential, this chapter argues that millets are key to achieving sustainable agricultural practices, improving rural livelihoods, and driving economic growth in a climate-stressed world. The findings aim to guide policymakers, researchers, and stakeholders in unlocking millets' full potential for a sustainable future.

Keywords: Climate-smart agriculture; SDG; millet; sustainable likelihood; value addition; millet trade

1. Introduction

Achieving SDGs in India plays an important role in eradicating poverty, ensuring inclusive growth, and leading a sustainable future for a growing and diverse population. As India is home to 17% of the global population, India's progress on SDGs significantly influences global outcomes (Pallayil et al., 2024). Therefore, transforming the agriculture sector which is the backbone of the Indian economy and supports nearly 50% of the population is crucial (Rajsinghot et al., 2024). India faces many agricultural challenges like fragmented landholdings, water scarcity, soil degradation, and climate change's adverse effects, which threaten food security and farmer's livelihoods (Sissoko et al., 2023). Therefore, transitioning to producing climate-smart products and renewable energy, enhancing irrigation efficiency, and promoting agroecology are critical for combating climate-related disruptions.

By integrating sustainable farming methods, empowering farmers, and improving supply chains, India not only secures food security domestically but also strengthens its role as a global agricultural leader, ensuring that development is inclusive and environmentally sustainable. With this context, CSA focuses on the techniques that enable climate resilience through sustainable practices like water-efficient irrigation, crop diversification, and precision farming. CSA is a strategic approach that aims to develop a farming system that can increase productivity sustainably with resilience to climate change and reduction of greenhouse gas emissions (Lata et al., 2013; Musafiri et al., 2020). Through CSA techniques, the livelihood of millions of small and marginal farmers can be protected from climate risks. Implementing CSA is vital in Indian agriculture to ensure long-term stability and target SDGs related to hunger, poverty, climate action, and sustainable use of resources (FAO, 2023).

"Millet" seamlessly harmonizes with the principles of CSA as they are resilient and nutrient-rich crops that are capable of thriving in adverse climatic conditions. These crops are ideal for addressing food security and sustainability challenges in India (Ajatasatru et al., 2024). These "super grains" are naturally drought tolerant and require significantly less water and chemical inputs as compared to staple crops like rice and wheat (Lenka et al., 2020). This helps in reducing the strain on critical resources and lowering greenhouse gas emissions. Their ability to grow on marginal lands with minimal inputs makes them a sustainable option for small and marginal farmers, enhancing agricultural resilience against climate variability. By incorporating millets into farming systems, India can advance CSA objectives, ensuring sustainable production, climate adaptability, and improved rural livelihoods while meeting the growing demand for environmentally friendly and health-conscious food options (Ahmed et al., 2024).

This chapter studies the present research work on CSA at a global level and attempts to show how the principles of CSA and SDGs completely harmonize together. The goals of sustainable development cannot be achieved without following a climate-resilient agriculture system through CSA techniques. This chapter studies millet as a crop that completely aligns with the SDGs, such as Zero Hunger, Good Health and Well-being, Decent Work and Economic Growth, and Climate Action. This chapter also explores millet as a sustainable livelihood alternative for marginalized farmers, especially women and evaluates the trade potential of India by introducing an experiment in the current scenario.

This chapter therefore explores the multifaceted potential of millets in India, aligning with the SDGs, particularly SDG 2 (Zero Hunger), SDG 3 (Good Health and Well-being), SDG 13 (Climate Action), and SDG 8 (Decent Work and Economic Growth). This chapter in subsequent sections highlights the transformative role of value addition in millet-based enterprises, focusing on women-led initiatives that enhance income and promote gender equity. The study also outlines how innovative millet products and improved market linkages bolster rural economies while catering to the rising demand for health-conscious foods globally. Lastly, this chapter examines the potential of the millet industry, emphasizing India's potential to position itself as a leader in the global millet market, through general equilibrium analysis, studying the associated spillover effects and complexities. By integrating millets into CSA frameworks, fostering value chain development, and scaling their trade potential, this chapter argues that millets are key to achieving sustainable agricultural practices, improving rural livelihoods, and driving economic growth in a climate-stressed world. The findings aim to guide policymakers, researchers, and stakeholders in unlocking millets' full potential for a sustainable future.

2. Literature Review on CSA and Millets

To review the existing literature, Scopus database was accessed, and data were extracted using the link

> Title-Abs-Key (Climate And Smart And Agriculture) And Title-Abs-Key ("Millet" Or "Millets")) And (Limit-To (Subjarea, "Econ") Or Limit-To (Subjarea, "Soci")) And (Limit-To (Doctype, "Ar") Or Limit-To (Doctype, "Ch")) And (Limit-To (Pubstage, "Final")) And (Limit-To (Srctype, "J") Or Limit-To (Srctype, "B") Or Limit-To (Srctype, "K")) And (Limit-To (Language, "English").

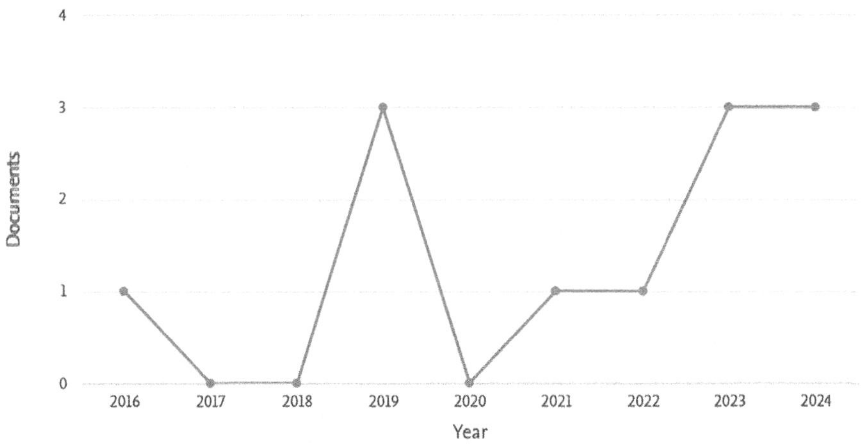

Fig. 13.1. Analysis of Documents by Year Under CSA and Millet/Millets.
Source: Scopus database.

The analysis of existing research (Fig. 13.1) clearly depicts that the research on millets is gaining momentum in recent times.

Millets are particularly significant in developing countries, where smallholder farmers face growing risks due to climate variability. Their low input requirements and ability to thrive in degraded soils make them a sustainable alternative to water-intensive staples like rice and wheat. Furthermore, the analysis of existing literature (Table 13.1) clearly shows the rising global interest in adopting millets as part of climate-adaptive strategies to combat food insecurity, promote biodiversity, and support rural livelihoods. Thus, the intersection of millets and CSA emerges as a vital research area for addressing the dual challenges of climate change and global food demand.

Table 13.1. Existing Literature on CSA and Millets (Year Wise).

Title of the Paper	Year
Can more drought resistant crops promote more climate secure agriculture? Prospects and challenges of millet cultivation in Ananthapur, Andhra Pradesh	2016
Agro-met services and farmer responsiveness to advisories: implications for climate smart agriculture	2019
Impact of smart crop-livestock diversification as climate change adaptation strategies on farmers' living conditions, Tahoua State, Niger Republic	2019
Assessment of agricultural activities and market for local crops under changing climate conditions in Gujarat	2019

Table 13.1. (*continued*)

Title of the Paper	Year
Contribution of climate-smart agriculture technologies to food self-sufficiency of smallholder households in Mali	2021
Investigating the pearl millet (*Pennisetum glaucum*) as a climate-smart drought-tolerant crop under Jordanian arid environments	2022
Mobile technology for smart agriculture: deployment case for pearl millet disease detection	2023
Climate-smart agriculture: improving dryland crop yields and value addition through university-community partnership in Zimbabwe	2023
Economic empowerment of women, millet farming, and sustainable development	2023
Does climate change affect crops differently? New evidence from Nigeria	2024
Saving farm subsidies with smart climate interventions: the case of transition to a millet-based agriculture	2024
Economy-wide impact of climate smart agriculture in India: a SAM framework	2024

Source: Scopus database.

Insightful but limited research done on Indian agriculture where Ajatasatru et al. (2024) observe that alternative crops like millet show minimal gains in income and output, and they offer significant environmental advantages over rice and wheat. The study suggests that these results could inform policy decisions to scale up CSA interventions in India to improve sustainability and food security. Also, recent literature by (Sedithippa Janarthanan, 2024) assesses how transitioning to millet-based agriculture can optimize farm subsidies and improve sustainability. It presents millet as a cost-effective alternative under climate change scenarios.

Medhekar (2023) in her study focuses on women-led projects that increase revenue and advance gender parity, highlighting the transformative power of value addition in millet-based businesses. Additionally, the report describes how new millet products and enhanced market connections support rural economies and meet the growing demand for health-conscious diets around the world.

The literature review analysis shows the rising scope of studying millets with reference to CSA and SDG practices where almost 37.2% of studies done accrue to the field of Economics and Social Sciences (refer Table 13.2). This study therefore builds on existing literature and tries to add meaningfully on how millet production can play a significant role in achieving various SDGs through CSA practices with reference to the Indian economy.

Table 13.2. Analysis of Documents by Subject Area.

Category	Value (%)
Economics, econometrics, and finance	18.6
Social sciences	18.6
Engineering	14.0
Environmental science	14.0
Agricultural and biological sciences	11.6
Business, management, and accounting	7.0
Computer science	7.0
Energy	4.7
Decision science	2.3

Source: Scopus database (authors').

3. Millet Production and SDGs

The principles of SDG align perfectly with the principles of CSA, by adopting agriculture practices that are climate resilient, enhancing food security, lower water footprint, and enhancing crop diversification. Millets, also known as "super grains," are becoming increasingly important in enhancing sustainable lifestyles, especially in rural and semi-arid areas where agriculture is greatly impacted by environmental issues. Because of their innate resistance to climate change, these crops are ideal for growing in regions that frequently experience unpredictable rainfall and protracted droughts. Millets need a lot less water than crops like rice and wheat, and they may grow in marginal soils with little help from outside sources. By lowering the risks of crop failure and guaranteeing consistent production, this adaptability gives smallholder farmers a source of reliable income, who frequently depend on agriculture for their sustenance.

Table 13.3 highlights the production of major crops across Indian states, including rice, wheat, maize, and nutri/coarse cereals (which majorly include millet), along with their contribution to the country's total production. Uttar Pradesh emerges as the leading producer of total food grains, contributing 18.03%, followed by Madhya Pradesh and Punjab. Major constituents of the total food grains accruing to Uttar Pradesh, Punjab, and Madhya Pradesh constitute wheat and rice production. Such agricultural practices are not sustainable in the long term, especially in water-scarce regions. Therefore, a shift toward cultivating less water-intensive crops, like millets, can be a feasible solution to make agricultural practices more sustainable and climate smart.

With regard to 2023 being the International Year of Millets (FAO, 2023), millets help in targeting three SDGs. It is directly connected with SDG 2 (Zero Hunger) because growing and eating millet gives a healthy and stable source of food. Moreover, the millet agriculture supports SDG 8 (Decent Work and Economic Growth) by providing livelihood to the smallholder farmers. Being

Table 13.3. Largest Food Grains Producing States of India 2023–2024.

Crops/Groups of Crops	States	Production (Million Tons)	Percentage Share of Production to All India
Food grains			
Rice	Telangana	16.63	12.17
	Uttar Pradesh	15.72	11.5
	West Bengal	15.12	11.06
Wheat	Uttar Pradesh	35.43	31.38
	Madhya Pradesh	21.28	18.84
	Punjab	17.78	15.75
Maize	Karnataka	5.49	15.39
	Bihar	4.61	12.93
	Madhya Pradesh	4.33	12.14
Total nutri/coarse cereals	Rajasthan	8.03	14.66
	Karnataka	7.61	13.9
	Madhya Pradesh	5.49	10.02
Total food grains	Uttar Pradesh	59.29	18.03
	Madhya Pradesh	39.84	12.12
	Punjab	32.59	9.91

Source: ES&E Division, Department of Agriculture and Farmers Welfare website: https://agriwelfare.gov.in/en/Agricultural_Statistics_at_a_Glance.

a resistant and drought-tolerant crop, millet helps to achieve SDG 13 (Climate Action) as it can survive in the harsh growing environment with minimum water requirements. Its nutritional content, which contains the necessary minerals and fibre, facilitates SDG 3 (Good Health and Well-Being) by preventing malnutrition and diet-associated diseases (Babele et al., 2022).

There is a demand globally to increase the harvesting of C4 cereals, including millets, and reduce the cultivation of C3 crops such as rice and wheat (Wambi et al., 2021). This is because millet cultivation is a more climate-sustainable (SDG 13) option for farmers as it requires less water, fertilizers, and pesticides as compared to water intensive crops like rice and wheat (Derbile et al., 2022). Millets are climate-resilient crops that can withstand high temperatures and prolonged periods of drought. Their ability to thrive in adverse conditions makes them crucial in the fight against climate change. By integrating millet cultivation into farming systems, countries like India can reduce the vulnerability of their agricultural sectors (Bandyopadhyay et al., 2017).

Despite being historically labeled as "poor man's food grains," millets are viable and beneficial staples for improving health and sustainability, which can be advocated with wider adoption in modern diets (Ghosh, 2024). Millet grains are rich in protein, dietary fiber, vitamins, and minerals such as iron, calcium, and magnesium, making them a vital source of nutrition for communities facing food insecurity (SDG 2). Increasing the consumption of millets could contribute to food security in India and help meet the growing demand for nutritious, climate-resilient food (Yiridomoh et al., 2024).

Furthermore, millets are considered a superfood due to their health benefits. Millets are presented as nutrient-dense grains, rich in proteins, essential amino acids, dietary fiber, and antioxidants (Anal et al., 2024). Studies show that millets can help overcome certain lifestyle diseases such as diabetes, obesity, and cardiovascular issues. Regular consumption of millet can also reduce the risk of anemia, particularly in women and children, as they are rich in iron and other micronutrients. Promoting millet-based diets through including millets in snacks, and baby food can help reduce the burden on healthcare systems (Diksha Bhatt, 2023).

Also, as a climate-resilient crop, millets have the potential to significantly increase rural incomes in India (SDG 8). These small-seeded cereals, which include types like Foxtail millet (Kangni), Pearl millet (Bajra), and finger millet (Ragi), are becoming more and more praised for their nutritional content, capacity to withstand severe growing circumstances, and resilience to climate change (Nambiar et al., 2020). In India, where agriculture provides for about half of the population, millets are becoming a "Super Crop" for CSA as well as a viable way to generate income in rural areas, particularly for women.

4. Role of Millet in Sustainable Livelihood and Global Trade

4.1. Millet and Sustainable Livelihood

Millets are rapidly being accepted by CSA in India as a source that provides farmers with improved livelihood opportunities through market diversification and value enhancement (Traore et al., 2021). Additionally, millets, which use 30%–40% less water than wheat and rice and are known for their climate resistance, are becoming a reliable source of income even in harsh weather circumstances (Nambiar et al., 2020). Millets, which are staple crops in areas like Rajasthan, Karnataka, and Odisha, not only help with CSA but also lessen farmers' reliance on erratic rainfall and guarantee a steady output. Similarly, by lowering irrigation costs and raising market demand for millet-based products, government-backed initiatives supporting millet production in Karnataka under the Krishi Bhagya scheme have doubled farmer incomes (Government of Karnataka, 2022). In turn, this has given small farmers a steady source of income, even during protracted drought seasons.

Furthermore, across Indian states, women play a central role in millet cultivation, processing, and value addition (Rao & Reddy, 2019). By engaging in millet production women are not only gaining access to economic opportunities that are empowering them beyond subsistence farming but also in some cases creating scope for entrepreneurship. In Telangana, Deccan Development Society (DDS) has been a pioneer in promoting women-led millet farming. Women farmers, predominantly from marginalized communities, cultivate millets using organic practices. The DDS supports them by facilitating direct sales through farmer cooperatives, bypassing intermediaries, and as a result, women have witnessed an income growth of 40% compared to traditional paddy farming (Rao & Reddy, 2019).

With increasing global demand for sustainable, health-conscious food products, millets are gaining attention in urban markets, both domestically and internationally. Investments in millet-based value chains, including processing, packaging, and branding, can create employment opportunities and boost rural economies (Rao & Reddy, 2019). The introduction of millet-based value-added products can be instrumental in transforming rural economies. Products like millet cookies, flour, and ready-to-eat snacks which are witnessing increased demand, both domestically and internationally can play a pivotal role to empower women by engaging them in processing and small-scale manufacturing (Lokur et al., 2023).

Millet-processing industries, which require relatively low capital investment, can empower local communities, particularly women and small entrepreneurs, by generating additional income sources. In Odisha's Koraput district, the Mission Shakti initiative empowered women Self-Help Group (SHGs) to produce and market millet-based products like biscuits and snacks. Supported by the Odisha Millet Mission (OMM), women received training in processing, packaging, and branding, allowing them to access urban markets. Income from these initiatives has increased household earnings by 25%–30%, enabling families to invest in education and healthcare (Sahoo et al., 2024). In Madhya Pradesh's Mandla district, tribal women cultivating kodo and kutki millets under a government-supported initiative reported dual benefits: surplus production was sold for profit, while consumption improved family nutrition. Many households saw a reduction in anaemia rates among women and children, as documented by local nongovernmental organizations (NGOs) (Sinha et al., 2022).

The Indian government has been actively promoting millets through initiatives like the National Year of Millets (2018) and International Year of Millets (2023). Schemes such as the Rashtriya Krishi Vikas Yojana and the OMM have helped establish market linkages for farmers. Women farmers have particularly benefitted from these programs, with targeted subsidies and training aimed at enhancing their productivity and entrepreneurial skills. Despite their benefits, millet farming faces challenges such as limited market access, inadequate storage infrastructure, and competition from subsidized rice and wheat under the Public Distribution System. Additionally, cultural perceptions that millets are "poor man's food" hinder widespread adoption in urban markets.

Governments must encourage millet cultivation through improved pricing strategies, storage facilities, and export prospects to remove these obstacles. To ensure demand-driven growth, NGOs and business entities should help raise awareness about the health benefits of millets. Targeted interventions such as direct market access, microfinance assistance, and skill-building initiatives are essential for women. Women can increase their negotiating power and guarantee fair income distribution by establishing cooperatives and farmer-producer organizations under their leadership. Millets have the potential to propel an inclusive and sustainable agricultural economy in India with the help of market connections, community-driven initiatives, and supportive legislation.

4.2. High Trade Potential of Millets

The global acceptance of millets has been increasingly evidenced through the rise in its trade. The same has been rising exponentially except in the years 2020 and 2021, possibly due to COVID-19. Fig. 13.2 shows the trend of millet exports globally in the last seven years. Apart from the two years of pandemic, the overall trade of millet in the world has been on the rise, and total millet exports have been valued at approximately US$52,402.544 (US$1,000) in 2023. International trade of millet shows a great potential to grow in the future with the help of proper infrastructure. With the enhancement of CSA techniques, countries that adopt such crops can benefit by increasing production for trade purposes (Mandumbu et al., 2023).

Increasing the global trade of millet is a promising alternative to enhancing millet production in India. The government is actively engaging in research and development to boost millet production and emphasizing it as the food of choice.

Table 13.4 represents the largest millet exporting destinations of India in the last five years. The UAE being the prime destination is followed by Saudi Arabia. In the third position is Nepal even though it imposes the highest import tariffs on millet imports from India among these top destinations. Since Nepal is among the top 30 target millet export destination countries developed by the government, it will be more rewarding to the Indian agriculture sector if these import tariffs are reduced or lessened. Apart from Nepal, Sub-Saharan African (SSA) countries also impose high tariffs on Indian millet imports.

In this study, an experiment has been conducted by reducing the existing import tariff imposed by Nepal and SSA countries by 50%. Also, given the

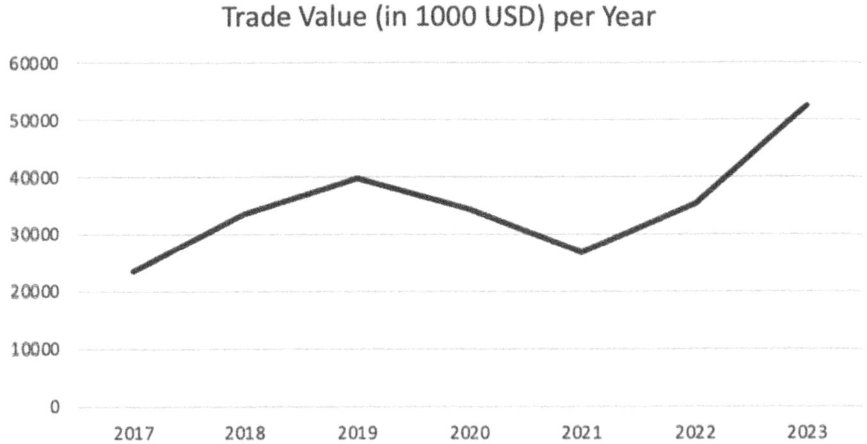

Fig. 13.2. Millet World Export (2017–2023). *Source*: UN Comtrade website.

Table 13.4. Largest Millet Export Destinations of India (in Million US Dollars).

Country	2019–2020	2020–2021	2021–2022	2022–2023	2023–2024
United Arab Emirates (UAE)	7.82	11.63	11.26	13.33	12.03
Saudi Arab	4.62	6.12	6.62	10.39	6.49
Nepal	4.33	6.55	5.94	5.57	6.11
United States	3.18	3.07	4.4	3.1	4.22
Senegal	0	0.02	0	1.14	3.71

Source: Directorate General of Commercial Intelligence and Statistics (DGCIS).

efforts of the Government of India to increase the production of millets, a 1% increase in factor productivity (land, labor, and capital) has been introduced as a simultaneous shock in the model. This study employs computable general equilibrium (CGE) analysis to address the stated objectives, utilizing the Global Trade Analysis Project (GTAP10P1) database for analysis and evaluation. The GTAP database is a comprehensive, publicly accessible global resource that documents bilateral trade flows, production, consumption, and intermediate use of goods and services, effectively modeling the global economy (Burfisher, 2021). Since its launch in 1993, GTAP has swiftly become a widely used tool for international economic analysis (Burfisher, 2021). For example, at the 1995 conference assessing the World Trade Organization's (WTO) Uruguay Round Agreement, three out of five quantitative analyses relied on GTAP data. Similarly, during the 1999 WTO and World Bank-sponsored Millennium Round of Multilateral Trade Talks in Geneva, nearly all global economic analyses were grounded in the GTAP framework. Furthermore, GTAP's significant role in recent Intergovernmental Panel on Climate Change meetings underscores its influence on economic research (Burfisher, 2021).

The model applied in this study is a standard static CGE model with a multi-region, multi-sector structure, operating under the assumptions of perfect competition, full employment, and constant returns to scale. Key features of the model include the representation of private household preferences using the non-homothetic constant difference of elasticities functional form, the explicit treatment of trade and transport margins in international trade, bilateral trade modeling via the Armington assumption, and the inclusion of a global banking sector that intermediates between global savings and consumption (Hertel, 1997).

The foundational model, comprising 141 regions and 57 sectors, has been adapted and restructured to align with the requirements of this study. The aggregation of regions and sectors was facilitated using the GTAP Aggregator, a tool developed by Mark Horridge. As part of this process, the regions in the model were reaggregated as follows: India, UAE, Nepal, Saudi Arabia, Oceania, East Asia, South East Asia, South Asia, North America, Latin America, European

Union (EU), Middle East and North Africa, SSA, and rest of the world. The sectoral aggregators include the following: millets, agriculture and allied activities, manufacturing, and services sector. In this study, one experiment was conducted, as stated below. RunGTAP software is used to introduce certain shocks in the model and the extracted results and then interpreted.

4.2.1. Experiment Results and Discussion

The experiment simulation of the reduction of import tariff on millets by in the trade of millets among SSA nations and Nepal along with a 1% rise in factor productivity projects a rise in India's gross domestic product by 0.003%. This will amount to an increase of a projected total of US$68.25 million. This depicts a positive outlook for India from this change. However, there is a notable observation (Table 13.5) due to these policy changes wherein industrial production of millets, agriculture and allied, and service sectors is rising; but that of the manufacturing sector is falling. The possible reallocation of resources responding to such government measures can lead to this situation. Nevertheless, it can be observed from Table 13.5 that overall industrial production is evaluated to receive a boost of nearly US$101.9 million.

Moreover, it was also observed that the supply price of commodities in different sectors will change with the above experiment. The results vary in different sectors. Particularly for millets and agriculture and allied sectors, the supply price is observed to reduce by 0.727% and 0.005%, respectively in India. Whereas for the manufacturing and services sector, the supply price is expected to increase by 0.002% and 0.005%, respectively. Rise in factor productivity leading to increased production can have a likely impact on supply price to fall. However, this can clearly lead to resource reallocation to harbor an inflated price in other sectors of the nation. Further, this possibly has an impact on India's international trade, as can be viewed from Table 13.5.

It is noteworthy that evaluations in Table 13.6 depict a percentage rise in India's exports of millets and agri-allied products to all the regions of the world. Strikingly, the same is seen to suffer in the case of manufacturing and services exports.

Table 13.5. Change in India's Industrial Production.

Sectors	**% Change**	**Change (US$ Million)**
Millets industry	0.239	30.341
Agriculture and allied industry	0.009	54.75
Manufacturing industry	−0.001	−17.375
Services industry	0.002	34.25

Source: Authors' calculations.

Table 13.6. Percentage Change in India's Exports.

Sector/Region	**UAE**	**Nepal**	**Saudi Arabia**	**United States**	**Oceania**	**East Asia**	**South-east Asia**	**South Asia**	**North America**	**Latin America**	**EU**	**Middle East and North Africa (MENA)**	**SSA**	**Rest of the World**
Millets	1.506	3.645	1.764	1.867	1.859	1.748	1.471	1.01	1.721	1.857	1.838	1.776	17.272	1.832
Agriculture and allied sector	0.036	0.268	0.034	0.04	0.039	0.038	0.039	0.034	0.04	0.04	0.04	0.038	0.038	0.041
Manufacturing	−0.011	0.049	−0.012	−0.013	−0.013	−0.012	−0.012	−0.011	−0.013	−0.013	−0.012	−0.012	−0.012	−0.013
Services	−0.019	0.206	−0.02	−0.019	−0.02	−0.019	−0.018	−0.018	−0.019	−0.02	−0.019	−0.019	−0.02	−0.019

Source: Authors' calculations.

Notably, the values of India's total exports are expected to fall as per the evaluations in Table 13.7. The significance of general equilibrium analysis is hence highlighted. It is clear that if the government makes efforts to enhance the production and exports of millet, it will have a positive impact on this industry and related economic variables. However, other sectors particularly in this case India's overall exports are found to suffer. Thus, the Indian government's steps toward Make in India for turning India into a global manufacturer becomes pertinent. Also, given the above experiment, the results found India's welfare to rise by US$74.5 million and global welfare by US$71.5 million.

5. Conclusion

The Indian government advocates millet as "super grain" and has been actively propagating the nutritional and environmental benefits of millet. This chapter aims to create more awareness about millet as a climate-smart crop perfecting aligning with SDGs which can also help in generating opportunities for the marginalized farmers. This potential can only be tapped when certain policy changes, and measures are taken to influence farmers to shift toward millet rather than water-intensive crops such as rice and wheat. Millets being drought resistant, require less water and are nutritionally superior, making them a sustainable alternative. States like Punjab, Haryana, and Uttar Pradesh, which currently focus on water-intensive crops, should consider adopting millets to ensure sustainable agricultural practices. Additionally, promoting millets aligns with India's goals of enhancing climate-resilient agriculture while addressing nutritional needs. Moreover, the rise in world trade of millets indicates the trend toward its global acceptance. Using general equilibrium analysis through GTAP, this study attempted to analyze the rise in factor productivity and partial reduction in Nepal's and SSA import tariffs on millet exports from India, on trade and industrial production. The situation was found to be favorable for the millet industry and agriculture and allied industries in terms of exports and industrial production. However, this can come at the cost of manufacturing and service production activity, which would require adjoining conducive policies for these sectors.

Table 13.7. Change in India's Exports (US$ Million).

Sectors	**Values**
Millets	19.1
Agriculture and allied sectors	12.9
Manufacturing	−30.9
Services	−19.9
Total	−18.9

Source: Authors' calculations.

References

Ahmed, J. S., Mahawar, S., & Mitra, S. (2024). Climate-smart agriculture – A potential way to reduce risks in agricultural ecosystems. In *Disaster risk reduction and rural resilience: With a focus on agriculture, water, gender and technology* (pp. 31–54). Springer Nature.

Ajatasatru, A., Prabhu, V., Pal, B. D., & Mukhopadhyay, K. (2024). Economy-wide impact of climate smart agriculture in India: A SAM framework. *Journal of Economic Structures*, *13*(1), 4. https://doi.org/10.1186/s40008-024-00252-w

Anal, A. K., Singh, R., Rice, D. L., Pongtong, K., Hazarika, U., Trivedi, D., & Karki, S. (2024). Millets as supergrain: A holistic approach for sustainable and healthy food product production. *Sustainable Food Technology*.

Babele, P. K., Kudapa, H., Singh, Y., Varshney, R. K., & Kumar, A. (2022). Mainstreaming orphan millets for advancing climate smart agriculture to secure nutrition and health. *Frontiers in Plant Science*, *13*, 902536. https://doi.org/10.3389/fpls.2022.902536

Bandyopadhyay, T., Muthamilarasan, M., & Prasad, M. (2017). Millets for next generation climate-smart agriculture. *Frontiers in Plant Science*, *8*. https://doi.org/10.3389/fpls.2017.01266

Derbile, E. K., Bonye, S. Z., & Yiridomoh, G. Y. (2022). Mapping vulnerability of smallholder agriculture in Africa: Vulnerability assessment of food crop farming and climate change adaptation in Ghana. *Environmental Challenges*, *8*. https://doi.org/10.1016/j.envc.2022.100537.

Ghosh, S. (2024). Current approaches and future potential for delivering CRISPR/Cas components in oilseeds and millets. *The Nucleus*, *67*, 141–156.

Government of Karnataka, Department of Agriculture. (2022). *State specific strategies for doubling farmers' income in Karnataka*. Indian Council of Agricultural Research (ICAR).

Lata, C., Gupta, S., & Prasad, M. (2013). Foxtail millet: A model crop for genetic and genomic studies in bioenergy grasses. *Critical Reviews in Biotechnology*, *33*(3), 328–343.

Lenka, B., Kulkarni, G. U., Moharana, A., Singh, A. P., Pradhan, G. S., & Muduli, L. (2020). Millets: Promising crops for climate-smart agriculture. *International Journal of Current Microbiology and Applied Sciences*, *9*(11), 656–668. https://doi.org/10.20546/ijcmas.2020.911.081

Lokur, A., Donde, K. J., & Pius, J. (Eds.). (2023). *MILLETS–2023: A transdisciplinary approach to its resurgence and sustainability*. Allied Publishers.

Mandumbu, R., Nyamadzawo, G., Nyakudya, W. I., Kamota, A., & Kubiku, F. N. M. (2023). Climate-smart agriculture: Improving dryland crop yields and value addition through university-community partnership in Zimbabwe. In A. Egeru, M. Lindow, & K. Muir-Leresche (Eds.), *University engagement with farming communities in Africa*. Community Action Research Platforms (pp. 226–236). https://doi.org/10.4324/9781003387497-19

Musafiri, C. M., Macharia, J. M., Ng'etich, O. K., Kiboi, M. N., Okeyo, J., Shisanya, C. A., Okwuosa, E. A., Mugendi, D. N., & Ngetich, F. K. (2020). Farming systems' typologies analysis to inform agricultural greenhouse gas emissions potential from smallholder rain-fed farms in Kenya. *Scientific African*, *8*. https://doi.org/10.1016/j.sciaf.2020.e00458.

Nambiar, V. S., Seshadri, S., & Shetty, P. (2020). Climate-resilient crops: Exploring the potential of millets in Indian agriculture. *Journal of Sustainable Agriculture*, *14*(3), 202–215.

Pallayil, B., Joseph, J., Veluthedan, S., Siddique, S. M. M., Bhaskaran, A. K., & Maheswary, U. (2024). India's sustainable development goal journey: Progress and pathways. In *Anticipating future business trends: Navigating artificial intelligence innovations: Volume 1* (pp. 423–437). Springer Nature.

Rajsinghot, K., Bala, S., & Singhal, P. (2024). Strengthening sustainable rural development through entrepreneurship: An Indian perspective. In *Informal economy and sustainable development goals: Ideas, interventions and challenges* (pp. 239–256). Emerald Publishing Limited.

Rao, D. S., & Reddy, P. N. (2019). Women in agriculture: Millet cultivation as a tool for economic empowerment. *Agricultural Economics Research Review*, *31*(1), 45–53.

Sahoo, M., Rana, K., & Samantray, P. K. (2024). Millet farming and tribal livelihood: An analysis of Odisha Millet Mission in Koraput District, Odisha, India. In *The role of women in cultivating sustainable societies through millets* (pp. 128–142). IGI Global.

Sanou, J., Bationo, B. A., Barry, S., Nabie, L. D., Bayala, J., & Zougmore, R. (2016). Combining soil fertilization, cropping systems and improved varieties to minimize climate risks on farming productivity in northern region of Burkina Faso. *Agriculture and Food Security*, *5*(1). https://doi.org/10.1186/s40066-016-0067-3.

Sedithippa Janarthanan, B. (2024). Saving farm subsidies with smart climate interventions: The case of transition to a millet-based agriculture.*China Agricultural Economic Review*, *16*(2), 249–266. https://doi.org/10.1108/CAER-05-2023-0129

Sikiru, A. B., John, M. O., Ambali, M. I., Muhammad, I. R., Egena, S. S. A., & Olatunbosun, R. B. (2023). Draught animal power for climate-smart farming in semi-arid and arid Nigeria: Challenges and opportunities. *Annals of Arid Zone*, *62*(3), 197–210. https://doi.org/10.59512/aaz.2023.62.3.2

Sinha, R., Sharma, K., & Das, M. (2022). Addressing rural malnutrition through millet farming: A case study from Madhya Pradesh. *Indian Journal of Nutrition Studies*, *28*(2), 85–98.

Sissoko, P., Guindo, S. S., Togola, S., Dembélé, B. D., Grimsby, L. K., & Aune, J. B. (2023). Effect of adoption of climate-smart-agriculture technologies on cereal production, food security and food diversity in Central Mali. *Agriculture (Switzerland)*, *13*(6). https://doi.org/10.3390/agriculture13061196

Traore, B., Birhanu, B. Z., Sangaré, S., Gumma, M. K., Tabo, R., & Whitbread, A. M. (2021). Contribution of climate-smart agriculture technologies to food self-sufficiency of smallholder households in Mali. *Sustainability (Switzerland)*, *13*(14). https://doi.org/10.3390/su13147757

Wambi, W., Otienno, G., Tumwesigye, W., & Mulumba, J. (2021). Genetic and genomic resources for finger millet improvement: Opportunities for advancing climate-smart agriculture. *Journal of Crop Improvement*, *35*(2), 204–233. https://doi.org/10.1080/15427528.2020.1808133

Yiridomoh, G. Y., Bonye, S. Z., & Derbile, E. K. (2024). Reducing vulnerability to climate change among millet and sorghum farmers in Ghana: Interrogating the contribution of climate-smart agriculture in northwestern Ghana. *Frontiers in Sustainable Food Systems*, *8*. https://doi.org/10.3389/fsufs.2024.1357355

Chapter 14

Ensuring Agricultural Sustainability: Evaluating Climate Policies and Their Impact

Kanhaiya Tripathi[a], *Deepak Kumar Chauhan*[b], *Kokila Lankathilake Konasinghe*[c] *and Naresh Prajapati*[b]

[a]*Dr Ambedkar Chair on Human Rights and Environmental Values, Central University of Punjab, Bathinda, India*
[b]*School of Legal Studies, Central University of Punjab, Bathinda, India*
[c]*Department of Public and International Law, Faculty of Law, University of Colombo, Sri Lanka*

Abstract

It is estimated that around one-fourth of the world's workforce is involved in agricultural operations, either directly or indirectly. Most of this labour is concentrated in countries with low and moderate incomes due to their geographical distribution and farming methods. Increasing temperatures, changing rainfall patterns, and severe weather events threaten food security and worsen economic inequality. As a result, many climate policies have been implemented worldwide to strengthen agricultural systems and encourage sustainable practices. These policies are crucial for protecting people's means of living and guaranteeing the long-term viability of agriculture. Nevertheless, discrepancies in these regulations highlight the need to comprehend the fundamental principles behind their development and implementation. This chapter thoroughly analyses current policies and programmes specifically designed to safeguard farmers against the adverse effects of climate change, with a particular focus on India. This chapter also examines the various ongoing cases and practices from India and other countries that rely on agriculture to assess the specific advantages these policies provide farmers and the safeguarding of their agricultural

Sustainable and Smart Farming Eco-systems:
Changing Landscapes in the Age of Technology, 199–217

Published under exclusive licence by Emerald Publishing Limited
doi:10.1108/978-1-83608-092-320261014

land and harvests. An analytical research methodology is utilised in this study, and secondary data sources, including academic literature, policy documents, and reports from international organisations, are employed as sources of information. A comprehensive content analysis was carried out to gain valuable insights into the policy creation process, the procedures involved in implementing policies, and the impact these policies have on agricultural practices. This chapter is concluded by proposing solutions to bolster the resilience of the agricultural sector.

Keywords: Agriculture; climate change; India; resilience; policies; sustainable practices

Introduction

The issue of climate change has emerged as a global preoccupation, impacting several countries. Madagascar is now experiencing food insecurity due to a severe drought, the world's first famine caused by climate change (United Nations, 2021). India saw a severe heatwave in 2024, which significantly negatively influenced human health (Mitra & Magramo, 2024). The 'Day Zero' disaster in South Africa brought attention to the acute water shortages in the country (Harding, 2021), while Bangalore, India, is also experiencing substantial water scarcity (Mogul, 2024). In addition, climate change has caused substantial flooding in Pakistan, greatly impacting the nation's overall socio-economic progress. Climate change adversely affects agriculture via unpredictable monsoons, droughts, and floods. This is seen in the irregular monsoons experienced in India (Pandey, 2023), the drought in the Horn of Africa (Paddison, 2023), and the flooding in Pakistan (Magramo, 2022). These changes in weather patterns negatively impact agricultural output and farmers' livelihood. As per the Food and Agriculture Organization's (FAO) *Statistical Yearbook World Food and Agriculture 2023* (n.d.), around 27% of the global workforce is employed in agriculture. The Department of Agriculture and Farmers Welfares states that 54.4% of the Indian workforce is employed in agriculture (Arun Kumar et al., 2022). Climate change has the potential to significantly decrease agricultural yield by 3%–7% for each one-degree increase in average temperature (Rezaei et al., 2023). This might lead to a higher likelihood of hunger and undernourishment. Most farmers in India, namely small and marginal farmers, who comprise about 86% of the overall number of farmers, have more significant difficulties caused by climate change than big farmers (Open Knowledge Repository, 2024). Large-scale farmers are more capable of adjusting to climate change, rendering small and marginalised farmers more susceptible.

Climate change considerably influences labour productivity in several economic sectors, including agriculture. Global warming might reduce around 5.8% in the total number of working hours, possibly eliminating over 34 million full-time employees, mostly concentrated in the construction and agriculture industries (Intergovernmental Panel on Climate Change (IPCC), 2021). The effects of climate change on agriculture varied across various areas in India. Projections indicate a rise in intense rainfall events in the western and south-central regions. Northwestern regions, such as dry western Rajasthan and nearby parts in Punjab and Haryana, are projected to have a decrease in summer monsoon rainfall and a rise in winter rainfall. This might result in increased occurrences of droughts and floods, impacting both the Kharif crops, which are normally planted between June and July and harvested between September and October, and the rabi crops, usually seeded between October and December and reaped between April and June. Climate change has the potential to significantly decrease agricultural output in India by 4.5%–9% per year from 2020 to 2039 (Saha et al., 2023). This would result in a yearly fall in the country's gross domestic product (GDP) of around 0.7%–1.35%. Appropriate climate change adaptation measures may reduce these problems (ET Online, 2024). The objective of this research is to cover aspects related to the impacts on agriculture directly and indirectly; further, the aim is to assess the effectiveness of existing climate policies and their influence on the sustainability of agriculture. The process includes evaluating the effectiveness of policies in mitigating agriculture's susceptibility to climate change, finding deficiencies, and offering suggestions to strengthen resilience via enhanced policy measures and sustainable practices.

Review of Literature

The study reviews existing literature on sustainable agriculture and climate change policy to identify key policy gaps and areas requiring further research. Saleem et al. (2024) examined securing a sustainable future: the climate change threat to agriculture, food security, and sustainable development goals (SDGs). The authors weigh the immediate impacts of climate changes, such as changes in temperature and rainfall, that pose a substantial risk to agricultural production and food systems. The indirect consequences include elevated food costs, disturbed distribution networks, and heightened susceptibility to pests and diseases. This study highlights the need to implement policies and calls for global and regional coordination to address the impacts of climate change, promote sustainable agriculture practices, and strengthen resilience to maintain long-term food security and economic steadiness. Maji and Selin (2024) studied a system analysis of the impact of agricultural policies on sustainability in India. The authors used the framework of the system to analyse the overall sustainability impacts of intervention in the case of the rice–wheat cropping system of Punjab. The authors examined the impact of air and water pollution effects on people's health in the region. This study suggested that either improving the existing cropping system or inducing fundamental changes in the cropping system can lead to substantial and wide-ranging sustainability benefits. Baraj et al.'s (2024) study delves into climate change and the resilience, adaptation, and sustainability of agriculture in India: a bibliometric analysis. The authors review the trends, patterns, and themes in research on climate change and the sustainability of agriculture in India. The authors analysed 572 publications published between 1994 and 2022 using bibliometric tools. This study focuses on climate change's local impact on agriculture, identifying important research areas and suggesting future policies to improve resilience and sustainability in the agricultural sector. Dey et al. (2023) scrutinised how climate-resilient agricultural ploys can improve livelihood and food security in eastern India. The authors examine the impact of climate-resilient tactics on paddy farmers in the Hooghly district of West Bengal. This study uses data from 612 farmers and probit regression, multinomial logistic regression, and propensity score matching to analyse the factors influencing adaptation and its results. The study's findings reveal that implementing measures such as crop insurance, drought-tolerant cultivars, and a system of rice intensification (SRI) have a substantial positive impact on paddy production, farmers' income, and food security. Deori et al. (2024) examined climate-resilient agriculture: enhancing agricultural sustainability in the face of climate change. The authors investigate the harmful effects of climate change on agriculture, such as rising temperatures, unpredictable rainfall patterns, and severe weather occurrences. This study emphasises the necessity to use vigorous agricultural systems against climate change to guarantee food security and sustainability on a global scale. This study offers a comprehensive plan for attaining agricultural sustainability by using technical breakthroughs, receiving regulatory backing, implementing climate-resilient practices, and highlighting the importance of agricultural scientists and policymakers in aiding the transition to a new system. Suresh et al. (2022) analysed agricultural sustainability and its trends in India: a micro-level index-based empirical evaluation. The authors study the level of agricultural sustainability in 17 Indian states between 1991 and 2011. The study utilises catalogues to measure sustainability's environmental, economic, and social aspects. This study establishes a general improvement in agricultural sustainability. However, notable regional differences and a negative correlation between economic and environmental sustainability indexes exist. The authors endorse an all-inclusive policy incorporating all three aspects to achieve sustainable agricultural expansion. Jatav and Naik (2023) explored measuring the agricultural sustainability of India: an application of the pressure–state–response (PSR) model. The authors employed the PSR model to evaluate the agricultural sustainability of 14 agro-climatic areas in India. This study produces a complete agricultural sustainability index using environmental, economic, and social limits. The Trans-Gangetic Plain Region is ranked the highest, mostly owing to variables such as high livestock ownership, concentrated agricultural practices, and access to established financing, and on the other side, the Eastern Himalayan Region is ranked the lowest, mainly because of its poor social and economic indices, despite its strong environmental sustainability. This study emphasises the need for comprehensive, geographically sustainable policies to improve agricultural sustainability. Streimikis and Baležentis (2020) examined the agricultural sustainability assessment framework, which integrates SDGs and interlinked environmental, climate, and agricultural policy priorities. This study incorporates the SDGs and environmental, climate, and agricultural policy priorities within the European Union (EU). This study conducts a thorough analysis of existing literature. It presents a new gauge framework that effectively integrates rural policy objectives with SDGs, climate change mitigation efforts, and environmental goals. The authors highlight the need to align different policies to achieve sustainable agriculture and emphasise the framework's potential for comparative evaluation across EU and member states. Ahmed et al. (2023) examined whether integrating climate change projection with agriculture policy improves agricultural adaptation and food security. This study highlights the need to coordinate agricultural policies with climate forecasts to reduce vulnerabilities caused by climate change. The study's findings reveal that there have been progressive rises in temperature and fluctuations in precipitation, resulting in the emergence of moderate to severe drought conditions. Nowak and Różańska-Boczula (2023) studied a comparative view of agricultural sustainability – the case of EU member states – using a combined measure incorporating economic, social, and environmental indicators. Based on these measures, the authors categorised the countries into four types, Austria, Estonia, Belgium, and Czechia, being identified as having the greatest level of sustainability. This study underlines the extensive variation in sustainability levels, accentuating the need for customised national policies. This study emphasises the need for ongoing surveillance and the adjustment of agricultural policy to accomplish equitable and continuing progress in various geographical zones.

The literature review emphasises the importance of sustainable agriculture and climate change strategies in guaranteeing food security and agricultural resilience. The studies emphasise climate change's direct and indirect effects on agriculture, the advantages of climate-resilient practices, and the need for spatially tailored policies. Although there has been substantial global and Indian progress, there are still pressing issues to address. Nevertheless, areas of study still need to be addressed to effectively combine localised climate projections with agricultural policy, tackle regional differences in sustainability, and improve the coordination of international and national policies.

Methodology

This study employs analytical methods to see the influence of sustainable agriculture and climate change policy on agricultural practices. The study seeks complete insights into policy formation, implementation methods, and their consequences on agricultural practices by using secondary data sources such as academic literature, policy papers, and reports from international organisations. Data collection entails a methodical examination of current scholarly literature to identify fundamental themes and areas that need further exploration. It also involves analysing policy documents and reports to comprehend the objectives and strategies.

Climate Change and Its Threats to Agriculture

Climate change substantially risks agriculture, affecting production, animal well-being, soil richness, and water availability. Rising worldwide temperatures result in updraft stress on crops and cattle, waning agricultural productivity. The

IPCC Report on Climate Change 2022: Impact, Adaptation and Vulnerability 2022 shows that yields of major cereal crops in climate-affected areas are already significantly lower than they were due to today's 1.1°C increase in global temperature average above pre-industrial levels. If that number reaches 1.5°C, the target set by IPCC in an earlier report as the highest we can go before a total climate disaster, about 8% of the world's farmland would become unsuitable for agriculture. As the current trajectory shows, an increase of 2°C or more could be catastrophic ('Climate Change Will Mean More Expensive Grocery Bills for Everyone, According to New IPCC Report', n.d.). Fluctuations in precipitation patterns, such as modified distribution of rainfall and heightened occurrences of droughts and floods, disturb agricultural operations. India has seen erratic monsoon patterns, leading to devastating droughts and floods that have impacted many farmers. In the past 40 years (1982–2022), 23% of districts, such as New Delhi, Bengaluru, Jaipur, etc., witnessed high numbers of efficient and excessive rainfall years (Prabhu et al., 2024). Severe weather occurrences like cyclones cause more frequent and intense damage to agricultural landscapes. The tropical cyclone Amphan largely affected agricultural land (78.2%) and forest, including mangroves (10.8%) in eastern India and Bangladesh (Kumar et al., 2021). Climate change impacts soil quality, increasing erosion, salinisation, and nutrient depletion. The Indian Council of Agricultural Research (Annual Report 2022–2023) conducted research emphasising how these variables lead to a decrease in soil fertility, which poses a danger to the practice of sustainable agriculture. Moreover, the proliferation of pests and diseases is facilitated by higher temperatures, hence augmenting the susceptibility of crops.

Food Security and Economic Inequality

Climate change significantly influences food security and economic inequality, which are crucial concerns at global and national levels. Around 80% of the global population is at the highest risk of crop failures. The greatest risks from climate change are found in Sub-Saharan Africa, South Asia, and Southeast Asia, where farming families are disproportionately poor and vulnerable (World Bank Group, 2022). As per the 2024 Global Report on Food Crises (GRFC, Global Network Against Food Crises), due to climate change, almost 72 million people in 18 countries faced high levels of food insecurity in 2023. Of 18 countries, 12 have 56.8 million acutely food-insured people needing urgent humanitarian assistance (Table 14.1) (*Food Security Information Network (FSIN) Joint Analysis for Better Decisions*, n.d.).

As per CNN 2023, in India, due to climate change, rice stock was ravaged, and rice was exported (Farooqui, 2023). India was rated 111th out of 125 countries in the Global Hunger Index (GHI) of 2023, scoring 28. This ranking underscores the persistent difficulties in achieving food security for the Indian people. Climate change deteriorates these difficulties by impacting agricultural production, livelihoods, and food approachability (GHI, 2023). On the other hand, the Indian prime minister at the International Conference of Agricultural Economists in 2024 claimed that India has become a food surplus country and is also working to

Table 14.1. Number of People Who Faced High Levels of Food Insecurity (2018–2023).

	2018	2019	2020	2021	2022	2023
Conflict/security	73.9 million	77.1 million	99.1 million	139.1 million	117.1 million	134.5 million
	21 countries	22 countries	23 countries	24 countries	19 countries	20 countries
Weather extremes	28.8 million	33.8 million	15.7 million	23.5 million	56.8 million	71.9 million
	26 countries	25 countries	15 countries	8 countries	12 countries	18 countries
Economic shocks	10.2 million	24.0 million	40.5 million	30.2 million	83.9 million	75.2 million
	6 countries	8 countries	17 countries	21 countries	27 countries	21 countries

Source: FSIN, GRFC 2019–2023 (https://www.fightfoodcrises.net/global-report-food-crises).

provide solutions for global food and nutritional security (MSN, 2024). Climate change is also escalating economic inequality; the Climate Inequality Report 2023 highlighted the concerns about economic inequality between countries regarding carbon emission distribution and energy consumption (Admin, 2023). World Bank report findings reveal that climate change will drive 68 million to 135 million into poverty by 2030. Climate change is a grave threat to South Asian countries, where most global poor are concentrated (World Bank, 2020). The research has found that the climate crisis in India could cost the country from 6.4% to more than 10% of its national income by 2100, taking 50 million more people back into poverty due to economic inequality (Mehta, 2023). The Grantham Research Institute on Climate Change and the Environment published a study that shows how climate change creates economic inequality in India due to extreme weather conditions like floods, which have cost India US$26.3 billion with damages exceedingly approximately 0.5% of its GDP (Choudhary, 2022). The Indian government launched many initiatives to tackle these issues, like the Public Distribution System, the National Food Security Act, and the Anatyodaya Anna Yojana. Still, the climate crisis, crop destruction, and low productivity triggered these initiatives and created an economic divide among the people (George & McKay, 2019). To summarise, climate change worsens the problem of food insecurity and economic inequality, especially in susceptible areas such as South Asia and Sub-Saharan Africa. India's efforts to address these problems highlight the pressing need for worldwide action to guarantee food security and decrease economic inequalities.

Global Climate Policies for Agriculture

Implementing global climate policies for agriculture is critical to reducing climate change's effects and assuring food production's long-term feasibility. International organisations and countries have formed and implemented many policies and programmes to tackle these difficulties. The United Nations Framework Convention on Climate Change (UNFCCC), 1992, was signed and enforced in 1994, and it is the first international convention to address climate change issues. Article 4 of the convention binds the state parties to take commitments into account common but differentiated responsibilities and their specific national and regional development and development priorities and promote cooperation with the relevant sectors, including agriculture, forestry, etc. (United Nations, 1992). The United Nations FAO supports member countries in advancing climate-smart farming practices. The climate-smart agriculture (CSA) method, implemented by the FAO, seeks to revolutionise agricultural systems by improving production, improving resilience, and minimising emissions (Food and Agriculture Organization of the United Nations, 2024). The Global Alliance for Climate-smart Agriculture (GACSA, n.d.), established by FAO, advocates for the worldwide implementation of climate-smart agricultural practices. The IPCC suggests scientific evaluations on climate change to assist in formulating policies. The IPCC's (2019) *Special Report on Climate Change and Land* emphasises the need to implement sustainable land management practices, specifically emphasising policies that support soil health, biodiversity, and water conservation in the agricultural sector. The World Bank and the International Fund for Agricultural Development (IFAD) provide funding for programmes that promote agriculture that can endure and adapt to climate change. The Climate-Smart Agriculture Investment Plan (CSAIP) framework, established by the World Bank in 2017, supports governments in identifying and prioritising investments in sustainable agriculture (World Bank, 2019). The Adaptation for Smallholder Agriculture Programme (ASAP; IFAD, n.d.), initiated by IFAD in 2012, seeks to enhance smallholder farmers' ability to withstand climate change impacts by making specific investments. The Global Environment Facility (GEF) supports initiatives that tackle the effects of climate change on agriculture. The GEF's Integrated Approach Pilots (2020) initiative seeks to enhance the sustainability of agricultural systems in Sub-Saharan Africa by focusing on improving soil health, water management, and agroforestry techniques.

Several global agricultural organisations also contribute to the practice of sustainable agriculture. The CGIAR's (2018) research programme on climate change, agriculture, and food security (CCAFS), initiated in 2011, aims to tackle the obstacles posed by climate change to guarantee food security. The Agrobiodiversity Index, developed by Bioversity International and introduced in 2017, quantifies and oversees the state of agrobiodiversity to enhance agricultural methods (The Agrobiodiversity Index|International Treaty on Plant Genetic Resources for Food and Agriculture|Food and Agriculture Organization of the United Nations, 2024). International Center for Agricultural Research in the Dry Areas (ICARDA's) Dryland Systems Programme, introduced in 2013, advocates for adopting sustainable agricultural practices (ICARDA, Communication Team, 2014).

The Indian prime minister launched the National Action Plan on Climate Change (NAPCC) on 30 June 2008. The NAPCC aims to enable the country to adapt to climate change and enhance the ecological sustainability of India's development path. The NAPCC mainly focuses on eight sustainability-related missions like the National Mission for Sustainable Agriculture (NMSA), Green India, etc. The NMSA was operationalised in 2014–2015 through various government initiatives and components. The NMSA aims to make agriculture more productive, sustainable, remunerative, and climate resilient by promoting location-specific integrated/composite farming systems. The NMSA component is Farm Water Management 2014–2015 to enhance water use efficiency, and these activities have been subsumed under the Per Drop More Crop and Pradhan Mantri Krishi Sinchayee Yojana. Soil Health Management aims to promote integrated nutrient management through judicious use of chemical fertilisers, including secondary and micronutrients, in conjunction with organic manures and biofertilisers to improve soil health and productivity. The Soil Health Card Scheme was implemented in 2015 to provide soil health cards to all farmers in the country. It will inform farmers of their soil's nutrient status and recommendations on the appropriate nutrient dosages for improving soil health and fertility (Ministry of Environment, Forest, and Climate Change, 2021).

Resilient Farming Systems

Resilient farming systems denote agricultural practices and institutions built to endure and adjust to different shocks and strains, such as climate change, economic instabilities, and societal burdens (Meuwissen et al., 2019). The objective of these systems is to uphold production and profitability while also guaranteeing sustainability and environmental well-being. The significance of resilient agricultural systems resides in their capacity to provide consistent food supply, safeguard natural resources, sustain farmer livelihoods, and contribute to climate change mitigation and adaptation (Chao, 2024). Encouraging sustainable practices is crucial for the establishment of resilient agricultural systems. These practices include a range of techniques such as varied agricultural systems, conservation agriculture, agroforestry, integrated pest control, efficient water use, and soil health enhancement. These methods bolster the ability of agricultural systems to withstand and recover from challenges by enhancing the variety of species, decreasing reliance on chemical inputs, and boosting the effectiveness of resource utilisation (Pantera et al., 2021).

Resilient farming systems' components are sustainable agricultural practices, crop diversification, soil health management, water conservation techniques, use of technology (drones, sensors), climate-resistant crop varieties, integrated pest management, and community-based support systems. Various initiatives and schemes were launched at global and national levels to make agriculture more resilient. The FAO's CSA, the World Bank's CSAIP, CGIAR's research programme on CCAFS, and IFAD's Adaptation for Smaller Agriculture programme are some of the important initiatives and schemes launched by international institutions to make agriculture climate resilient.

India has implemented many initiatives and schemes to promote resilient farming nationally. The Pradhan Mantri Fasal Bima Yojana (PMFBY, 2019), initiated in 2016, offers agricultural insurance to farmers, safeguarding their revenues from natural disasters. In the Union Budget for 2024–2025, INR 15,000 crore was allotted for the PMFBY. This allocation demonstrates the government's commitment to ensuring stable incomes for farmers in the face of climate-related hazards and making agriculture resilient for farmers (MSN, 2024). The Paramparagat Krishi Vikas Yojana (PKVY) encourages organic farming and sustainable techniques that improve soil health and biodiversity. The Pradhan Mantri Krishi Sinchai Yojana (PMKSY), launched in 2015, aims to enhance water utilisation efficiency by promoting micro-irrigation and water conservation (Agrifarm, 2024). The Rashtriya Krishi Vikas Yojana, initiated in 2007, assists state governments in executing environmentally friendly agricultural initiatives. The Soil Health Card Scheme, implemented in 2015, offers farmers' detailed information on the nutrient state of their soil and advises them on how to enhance it. This initiative aims to encourage sustainable land management practices. The Pradhan Mantri Kisan Samman Nidhi (PM-Kisan) plan offers financial assistance to farmers, with a budget of INR 60,000 crore for the fiscal year 2024–2025, demonstrating ongoing support for the agricultural community. This programme aims to provide immediate monetary support to small and marginalised farmers to improve their quality of life and decrease economic inequality. Notable allocations in the 2024–2025 budget include INR 955 crore for the establishment and promotion of 10,000 Farmer Producer Organisations (FPOs), INR 600 crore for the Agriculture Infrastructure Fund, and INR 500 crore for the NAMO Drone Didi initiative, which aims to modernise agriculture through technology. International and Indian efforts are vital to foster resilient farming systems that withstand and recover from shocks and stressors (*Highlights of the Union Budget 2024–25*, 2024). They also promote farmers' lives and improve agricultural systems' ability to adapt to different challenges.

Policy Effectiveness and Challenges

Enforcing worldwide climate policy for agriculture is essential for reducing the effects of climate change and guaranteeing the long-term viability of the food supply. Multiple global organisations and nations have formulated and executed strategies to tackle these difficulties. The UNFCCC, founded in 1992, was the first global agreement to tackle climate change. It fosters collaboration across several sectors, such as agriculture and forestry. The CSA approach developed by the FAO seeks to transform agricultural systems by increasing production, improving resilience, and reducing emissions. The GACSA promotes the worldwide implementation of CSA. The IPCC offers scientific assessments to guide policy decisions, focusing on promoting sustainable land management methods. The CSAIP by the World Bank aims to assist governments in allocating resources towards sustainable agriculture. Meanwhile, the ASAP by IFAD focuses on strengthening the ability of smallholder farmers to withstand and recover from challenges. The GEF and CGIAR's research programme on CCAFS support using environmentally friendly methods in agriculture.

The NAPCC was initiated in India in 2008 to facilitate the country's capacity to adjust to climate change and improve ecological sustainability. The NAPCC focuses on eight missions, including the NMSA, to foster productive, sustainable, and climate-resilient agricultural practices. The PMFBY is an initiative that offers agricultural insurance to safeguard farmers from natural calamities. On the other hand, the PKVY promotes organic farming. The PMKSY aims to enhance water utilisation efficiency. At the same time, the Soil Health Card Scheme offers farmers valuable insights on soil nutrients to encourage sustainable land administration. The agricultural community is further supported by financial assistance programmes such as the PM-Kisan and the distribution of funds for FPOs.

However, ongoing concerns need to be addressed, including gaps in implementing strategies, limitations in financing, difficulties in integrating policies, impediments to adopting new technologies, and insufficient monitoring and evaluation. Furthermore, the problem of farmer suicides underscores the profound socio-economic pressures experienced by the agricultural community. In 2019, the National Crime Records Bureau reported that more than 10,000 farmers and agricultural labourers committed suicide, and 557 farmers committed suicide in Maharashtra in the first six months of 2024 (TOI News Desk, 2024). These dreadful incidents were influenced by causes such as crop failure, debt, and insufficient help. To enhance the success of these programmes, it is crucial to reinforce implementation, provide more funds, ensure policy consistency, encourage technology usage, and improve monitoring and evaluation systems. Tackling these obstacles would greatly strengthen the capacity of agricultural systems to withstand and recover from the effects of climate change, guaranteeing food availability and maintaining economic stability.

Analysis of Various Ongoing Cases and Practices: India and Other Agrarian Economies

To ensure productivity, profitability, sustainability, and environmental health, resilient farming systems are agricultural techniques built to resist and respond to shocks like economic volatility and climate change. At the global level, small-scale farmers in the Brazilian Amazon have used agroforestry techniques, including combining products such as cacao, banana, and cassava with native and non-native trees. This has resulted in better soil fertility, heightened biodiversity, and improved carbon sequestration, which has led to increased incomes, diversified output, and enhanced resilience to market swings and climate change (Foote, 2019). In flood-prone areas of Bangladesh, farmers create buoyant platforms using water hyacinth and organic materials, on which they cultivate products such as vegetables and spices. Implementing these buoyant agricultural plots guarantees a dependable food supply during periods of heavy rainfall, enhances food stability, and mitigates the effects of climate change, thus improving the financial well-being and sustenance of farming communities (Sunder, 2020). Vietnamese farmers in the Mekong Delta have implemented the SRI by using techniques such as planting younger seedlings, spacing them at greater distances, and utilising organic fertilisers. This has resulted in higher rice yields, a reduction in water use by up to 30%, and a reduced dependence on chemical inputs. As a result, it has improved food security, enhanced farmer incomes, and promoted environmental sustainability (*Cornell Chronicle*, 2021). Communities in the Peruvian Andes have successfully revitalised and preserved old Andean terraces by combining traditional wisdom with contemporary agriculture methods. These efforts have enhanced the management of water resources, mitigated soil erosion, and boosted agricultural output. Additionally, they have successfully safeguarded cultural heritage and promoted sustainable lifestyles.

India has also implemented many effective programmes to enhance agricultural practices. These include the PKVY, which covers the country where farmers are provided financial assistance and organic inputs. Bhartiya Prakritik Krishi Padhati, a part of PKVY, promotes traditional indigenous practices among farmers. These schemes support organic farming in India. As a result of these schemes, Sikkim became the first Indian state full 100% organic farming state and set an example for the world in terms of best agriculture practices (Agri Farming, 2022). Andhra Pradesh will become India's first zero-budget natural farming state in 2024. The programme aims to eradicate synthetic chemicals, promote agricultural biodiversity, and increase farmers' earnings. The programme received accolades from international authorities for its capacity to accomplish

SDGs and revolutionise global farming methods. The initiative aims to provide US$2.3 billion over six years to support climate-resilient and regenerative agricultural practices. The objective is to safeguard local food systems and enhance the well-being of farmers (UN Environment, 2018). The SRI in Tamil Nadu, India, transforms rice cultivation by transplanting immature seedlings in grids far apart. This technique improves the development of plant roots, saves water by alternating between wet and dry conditions, and fosters a thriving soil ecology. SRI prioritises organic methods using compost and vermicompost as alternatives to artificial fertilisers, minimising its environmental footprint. SRI enhances crop productivity and adaptability to climate change while promoting sustainability and ensuring food security. This creative but traditional technique is revolutionising agriculture in Tamil Nadu, serving as a sustainable model for South Asia and beyond (Zachary, 2023). India also follows traditional farming practices that are environmentally sustainable, including mixed cropping, crop rotation, and agroforestry, which are essential for achieving sustainable farming. These techniques bolster soil health, diminish pest and disease cycles, and promote biodiversity. Utilising traditional seeds, implementing water-gathering techniques, and employing organic manuring methods enhance adaptability to local circumstances, provide dependable water resources, and enhance soil fertility. Methods such as mulching, vermiculture, and the use of Panchagavya enhance soil composition and promote plant development, hence decreasing the need for chemical substances. In addition, using bio-pesticides and zero tillage practices contribute to preserving ecological equilibrium and soil integrity. Practices such as Jhum cultivation, unique to Northeast India, contribute to the restoration of land and the preservation of biodiversity. These conventional techniques conserve India's farming legacy and foster ecological sustainability (Patel et al., 2020).

Integrating conventional and groundbreaking agricultural methods is crucial for guaranteeing efficiency, financial success, long-term viability, and ecological well-being in agrarian economies such as India and other countries worldwide. Successful types of sustainable agriculture, such as the agroforestry techniques in Brazil, buoyant farming in Bangladesh, the SRI in Vietnam, and terrace revitalisation in Peru, demonstrate the resilience of agricultural systems. India's initiatives, such as the PKVY, Zero-Budget Natural Farming in Andhra Pradesh, and the SRI method in Tamil Nadu, demonstrate successful approaches to improve agricultural sustainability. These techniques not only enhance the capacity of the environment to withstand and recover from disturbances but also enhance the ability of agricultural communities to support themselves, ensuring that they have enough food and economic stability.

Role of Insurance in Agriculture

Insurance is essential in agriculture since it offers farmers financial security against various hazards, including crop failure, natural catastrophes, pests, and illnesses. At the global and national levels, many insurance schemes have been launched to protect the interests of farmers and promote agricultural sustainability. The World Bank Group launched the Global Index Insurance

Facility (GIIF) to facilitate access to finance for smallholder farmers through catastrophic risk transfer solutions and index-based insurance in developing countries. The GIIF's (2024) regional partners have facilitated approximately US$2 billion in sum insured. The FAO (2011) launched the Agricultural Insurance in Asia and Pacific Region initiative and facilitated US$800 million for the Asian region in 2010 onwards. The Weather Risk Management Facility (WRMF) is a joint initiative by the World Food Programme and the IFAD established in 2008 to reduce smallholders' vulnerability to weather and other risks that limit agricultural production. It encourages and protects investment in smallholder agricultural production and enhances food security (WRMF|NDC Partnership, 2024). The R4 Rural Resilience Initiative was proposed by the World Food Programme and Oxfam America in 2011 to enable vulnerable rural families to increase their food and income security by managing climate-related risks. As of 2023, R4 had benefited nearly 550,000 vulnerable households and families in 18 countries across Africa, Asia, Latin America, and the Caribbean (World Food Programme, 2011). The Asian Development Bank (ADB) also launched Agriculture Insurance for Asian Countries (Mahul & Stutley, 2010). The National Agricultural Insurance Scheme, National Agricultural Insurance Scheme, and Modified National Agricultural Insurance Scheme were restructured as PMFBY in 2016. It is a prominent insurance scheme designed specifically for Indian farmers. The objective is to provide extensive protection against crop losses caused by extreme weather events. Since 2016, PMFBY has seen significant growth, with enrolment increasing by 27% in 2024 alone. Over 23.22 crore farmer applicants have received claims, amounting to INR 31,139 crore against a premium of INR 31.139 crore, providing INR 500 in claims for every INR 100 premium paid. PMFBY remains voluntary and has become more attractive to farmers, with 42% of non-loanee farmers enrolled in FY 2023–2024, enhancing financial stability amid natural calamities. In the 2024–2025 budget, the government allocated 14,600 crore for PMFBY (*Enrolment Under PMFBY Increase by 27% in Current Year so Far*, 2024). The Restructured Weather Based Crop Insurance Scheme (RWBCIS), implemented in 2007 and revamped in 2016, aims to alleviate the economic consequences of unfavourable weather conditions on agricultural produce. The system uses meteorological factors such as rainfall, temperature, and humidity as triggers for providing compensation to farmers who have been impacted by adverse weather conditions. As of 2023, the RWBCIS programme has successfully registered over 10 million farmers, providing them with prompt support amid adverse weather conditions (National Portal of India, 2024). The Unified Package Insurance Plan (UPIS) was implemented as a trial project in 2016–2017 in 45 districts from Kharif. A farmer (both loanee and non-loanee) can access a bank, whereas non-loanee farmers shall deliver through banks and insurance intermediaries (*Operational Guidelines Unified Package Insurance Scheme*, 2016). The Coconut Palm Insurance Scheme (CPIS), introduced in 2009, is a specialised programme to support coconut producers by offering financial aid if their coconut plants are affected by natural calamities, climatic risks, pests, and diseases. This scheme applies to healthy, nut-bearing coconut palms aged 4–60 years (7–60 years for tall varieties), and this scheme excludes losses due to theft, war, nuclear events, etc. The CPIS operates in all coconut-growing states through the Agriculture Insurance Company and state governments (Coconut Development Board (Ministry of Agriculture and Farmers Welfare, Government of India), 2024). Eventually, insurance is crucial in agriculture as it offers economic protection against a range of uncertainties, fosters the long-term viability of agricultural practices, and bolsters the stability of food supplies. Global initiatives such as the GIIF, WRMF, and R4 Rural Resilience Initiative, together with regional endeavours by the FAO and ADB, have greatly aided small-scale farmers. In India, programmes like PMFBY, RWBCIS, UPIS, and CPIS have played a crucial role in reducing the effects of natural disasters and climate-related hazards. These initiatives have seen significant expansion and acceptance, providing prompt assistance and economic stability to millions of farmers, thus promoting resilience and sustained agricultural progress.

Conclusions and Suggestions

Climate change presents a significant peril to agriculture, affecting productivity, livelihoods, and food security, particularly in countries with limited and moderate incomes. Rising temperatures, unpredictable precipitation patterns, and extreme weather events have greatly impacted global agricultural systems. This chapter analyses the efficacy of different climate policies and their influence on agricultural sustainability, specifically emphasising India and other countries that rely heavily on agriculture.

The study emphasises the need to implement climate-resilient agricultural practices and policies to safeguard farmers and promote sustainable agriculture. The implementation of global and national policies, such as the UNFCCC, CSA, IPCC recommendations, and India's NAPCC, has significantly tackled these difficulties. Nevertheless, ongoing challenges such as deficiencies in policy execution, financial constraints, and socio-economic burdens persistently hinder the progress of these endeavours.

Brazil, Bangladesh, Vietnam, Peru, and India case studies show successful examples of resilient agricultural systems and traditional practices that increase sustainability. In addition, crop insurance plans are essential in reducing financial risks related to climate change and fostering resilience.

To improve the execution of policies, it is crucial to develop the procedures for monitoring and evaluation, provide more funds and resources, and promote coherence and integration of policies across different sectors. We can greatly enhance resource efficiency and production by advocating for technical innovations such as precision farming, drones, and sensors and allocating resources towards research and development. Providing subsidies, training programmes, and inexpensive finance is crucial to support small and marginal farmers to enhance their resilience. Enhancing social safety nets and establishing community-based support structures may further bolster their ability to deal with climatic shocks.

It is essential to take measures such as increasing the scope of agricultural insurance, streamlining the process of making claims and enhancing farmers' understanding of the advantages of insurance. Advocating for sustainable agricultural techniques, including crop diversification, conservation agriculture, agroforestry, soil health management, and water conservation, is essential. Facilitating global cooperation and sharing information is crucial in tackling the worldwide scope of climate change and its effects on agriculture.

Governments, researchers, and agricultural stakeholders may adopt these recommendations to bolster the resilience of agricultural systems, safeguard farmers' livelihoods, and secure the sustainability of the global food supply. The insights and suggestions aim to promote sustainable agriculture and encourage the development of long-term food security.

References

Admin. (2023, April 15). *Climate inequality report 2023*. https://wealthcap.org/2023/04/15/climate-inequality-report-2023/

Agrifarm. (2024, July 31). *Paramparagat krishi vikas yojana (PKVY scheme)*. NIBSM https://www.nibsm.org.in/en/pkvyscheme/#:~:text=Paramparagat%20Krishi%20Vikas%20Yojana%20%28PKVY%29%2C%20i nitiated%20in%202015%2C

Agri Farming. (2022, September 28). *Organic farming in Sikkim: Crops grown, starting ideas for beginners*. https://www.agrifarming.in/organic-farming-in-sikkim-crops-grown-starting-ideas-for-beginners

Ahmed, Z., Ambinakudige, S., & Fosu, V. (2023). Does integrating climate change projection with agriculture policy improve agricultural adaptation and food security? Evidence from Bangladesh. *Theoretical and Applied Climatology*. https://doi.org/10.1007/s00704-023-04643-6

Arun Kumar, E., Sr, & Sharma, A.K., Yadav, R. K., Meitei, S., Arora, N., & Gaur, D. K. (2022). *Agricultural statistics at a glance 2022*. https://desagri.gov.in/wp-content/uploads/2023/05/Agricultural-Statistics-at-a-Glance-2022.pdf

Baraj, B., Mishra, M., Sudarsan, D., da Silva, R. M., & Santos, C. A. G. (2024). Climate change and resilience, adaptation, and sustainability of agriculture in India: A bibliometric review. *Heliyon, 10*(8), e29586–e29586. https://doi.org/10.1016/j.heliyon.2024.e29586

CGIAR. (2018). *Research program on climate change, agriculture and food security*. https://www.cgiar.org/research/program-platform/climate-change-agriculture-and-food-security/

Chao, K. (2024). Family farming in climate change: Strategies for resilient and sustainable food systems. *Heliyon, 10*(7), e28599–e28599. https://doi.org/10.1016/j.heliyon.2024.e28599

Choudhary, K. (2022, November 2). *How is India tackling climate change?* Grantham Research Institute on Climate Change and the Environment. https://www.lse.ac.uk/granthaminstitute/explainers/how-is-india-tackling-climate-change/

Climate change will mean more expensive grocery bills for everyone, according to new IPCC report. (n.d.). *Time*. https://time.com/6152615/ipcc-report-climate-change-agriculture/

Coconut Development Board (Ministry of Agriculture and Farmers Welfare, Government of India). (2024). *Coconut palm insurance scheme*. Retrieved August 19, 2024, from https://coconutboard.gov.in/docs/cpis-guidelines.pdf

Cornell Chronicle. (2021). *System of rice intensification recognised for climate policy impact*. https://news.cornell.edu/stories/2021/02/system-rice-intensification-re cognized-climate-policy-impact#:~:text=The%20System%20of%20Rice%20Intensification%20%28 SRI%29%20has%20been

Deori, C., Behera, P., Sharma, A., Kakoti, M., Das, J., Singh, A. P., Singh, B., & Das, R. (2024). Climate resilient agriculture: Enhancing agricultural sustainability in the face of climate change scenario. *International Journal of Environment and Climate Change, 14*(7), 394–410. https://doi.org/10.9734/ijecc/2024/v14i74279

Dey, S., Singh, P., Abbhishek, K., Singh, A. K., & Chander, G. (2023). Climate-resilient agricultural ploys can improve livelihood and food security in Eastern India. *Environment, Development and Sustainability*. https://doi.org/10.1007/s10668-023-03176-2

Enrolment Under PMFBY Increase by 27% in Current Year so Far. (2024). https://pib.gov.in/PressReleasePage.aspx?PRID=2011791

ET Online. (2024, June 24). India's FY24 CAD reduces to 0.7% of GDP from 2% in FY23, records surplus of 0.6% in Q4. *The Economic Times*. https://economictimes.indiatimes.com/news/economy/indicators/indias-cad-reduces-to-0-7-of-gdp-in-fy24-from-2-in-fy23/article show/111232715.cms

Farooqui, R. M. S. (2023, August 29). *Climate change has ravaged India's rice stock. Now its export ban could deepen a global food crisis*. CNN Business. https://edition.cnn.com/2023/08/28/business-india/india-rice-ban-export-farmers-loss-intl-hnk-dst/index.html

Food and Agriculture Organization of the United Nations. (2024). *Climate-smart agriculture*. https://www.fao.org/climate-smart-agriculture/en/

Food Security Information Network (FSIN)joint analysis for better decisions. (n.d.). https://www.fsinplatform.org/sites/default/files/resources/files/GRFC2024-full.pdf

Foote, N. (2019, October 17). *Agroforestry in the Amazon*. https://theecologist.org/2019/oct/17/agroforestry-amazon

George, N. A., & McKay, F. H. (2019). The public distribution system and food security in India. *International Journal of Environmental Research and Public Health, 16*(17), 3221. https://doi.org/10.3390/ijerph16173221

GIIF. (2024).*Overview*. https://www.indexinsuranceforum.org/overview

Global Alliance for Climate-smart Agriculture (GACSA). (n.d.). https://www.fao.org/gacsa/en/

Global Hunger Index. (2023). *The power of youth in shaping food systems*. https://www.globalhungerindex.org/pdf/en/2023.pdf

Harding, A. (2021, November 10). *Cape Town's day zero: "We are axing trees to save water."* BBC News. https://www.bbc.com/news/world-africa-59221823

Highlights of the union budget 2024–25. (2024). https://pib.gov.in/PressReleaseIframePage.aspx?PRID=2035609

ICARDA, Communication Team. (2014). *ICARDA annual report 2013*. https://www.icarda.org/annual-report-2013

IFAD. (n.d.). *Adaptation for smallholder agriculture programme*. https://www.ifad.org/en/asap

Integrated Approach Pilots. (2020). *Global environment facility*. https://www.thegef.org/what-we-do/topics/integrated-approach-pilots

Intergovernmental Panel on Climate Change (IPCC). (2019). *Special report on climate change and land*. IPCC. https://www.ipcc.ch/srccl/

Intergovernmental Panel on Climate Change (IPCC). (2021). *Climate change widespread, rapid, and intensifying*. IPCC. https://www.ipcc.ch/2021/08/09/ar6-wg1-20210809-pr/

Jatav, S. S., & Naik, K. (2023). Measuring the agricultural sustainability of India: An application of pressure-state-response (PSR) model. *Regional Sustainability, 4*(3), 218–234. https://doi.org/10.1016/j.regsus.2023.05.006

Kumar, S., Lal, P., & Kumar, A. (2021). *Influence of super cyclone "Amphan" in the Indian subcontinent amid COVID-19 pandemic. remote sensing in earth systems sciences*. https://doi.org/10.1007/s41976-021-00048-z

Magramo, K. (2022, September 2). *A third of Pakistan is underwater amid its worst floods in history. Here's what you need to know*. CNN. https://edition.cnn.com/2022/09/02/asia/pakistan-floods-climate-explainer-intl-hnk/index.html

Mahul, O., & Stutley, C. J. (2010). *Government support to agricultural insurance: Challenges and options for developing countries*. World Bank Group.

Maji, P., & Selin, N. E. (2024). A systems analysis of sustainability impacts of agricultural policies in India. *Earth's Future*, *12*(1). https://doi.org/10.1029/2023ef003667

Mehta, V. (2023, January 19). *India holds the key to hitting global climate change targets*. World Economic Forum. https://www.weforum.org/agenda/2023/01/india-holds-the-key-to-hitting-global -climate-change-targets-here-s-why/

Meuwissen, M. P. M., Feindt, P. H., Spiegel, A., Termeer, C. J. A. M., Mathijs, E., Mey, Y. de, Finger, R., Balmann, A., Wauters, E., Urquhart, J., Vigani, M., Zawalińska, K., Herrera, H., Nicholas-Davies, P., Hansson, H., Paas, W., Slijper, T., Coopmans, I., Vroege, W., & Ciechomska, A. (2019). A framework to assess the resilience of farming systems. *Agricultural Systems*, *176*, 102656. https://doi.org/10.1016/j.agsy.2019.102656

Ministry of Environment, Forest, and Climate Change. (2021). *National action plan on climate change (NAPCC)*. https://static.pib.gov.in/WriteReadData/specificdocs/documents/2021/dec/doc 202112101.pdf

Mitra, E., & Magramo, K. (2024, June 21). *"My body can't take it": Brutal nighttime temperatures give Delhi residents little respite from India's searing heatwave*. CNN. https://edition.cnn.com/2024/06/21/india/india-delhi-nighttime-heatwave-climate-intl-hnk/index.html

Mogul, R. (2024, March 15). *India's "Silicon Valley" is running dry as residents urged to take fewer showers and use disposable cutlery*. CNN. https://edition.cnn.com/2024/03/14/india/india-bangalore-water-crisis-impact-intl-hnk-dst/index.html

MSN. (2024). https://www.msn.com/en-in/foodanddrink/other/india-a-food-surplus-country-working-to-provide-solutions-for-global-food-security-pm-modi/ar-AA1oaaa6

National Portal of India. (2024). *Weather based crop insurance scheme (WBCIS)*. https://www.india.gov.in/weather-based-crop-insurance-scheme-wbcis

Nowak, A., & Różańska-Boczula, M. (2023). A comparative view of the level of agricultural sustainability – The case of European Union member states. *Sustainable Development*. https://doi.org/10.1002/sd.2804

Open Knowledge Repository. (2024). https://openknowledge.worldbank.org/bitstream/handle/10986/34726/Transforming-Rural-Farm-Livelihoods-The-NRLM-Journey.pdf? sequence=1

Operational guidelines unified package insurance scheme. (2016). https://pmfby.gov.in/pdf/FINAL_UPISOGs_23.03.2016.pdf

Paddison, L. (2023, April 27). *Catastrophic drought that's pushed millions into crisis made 100 times more likely by climate change, analysis finds*. CNN. https://edition.cnn.com/2023/04/27/africa/drought-horn-of-africa-climate-change-intl/index.html

Pandey, K. (2023, September 26). *Monsoon left widespread destruction and uneasy questions in Himachal*. https://india.mongabay.com/2023/09/monsoon-left-widespread-destruction-and-uneasy-questions-in-himachal/

Pantera, A., Mosquera-Losada, M. R., Herzog, F., & den Herder, M. (2021). Agroforestry and the environment. *Agroforestry Systems*, *95*(5), 767–774. https://doi.org/10.1007/s10457-021-00640-8

Patel, S. K., Sharma, A., & Singh, G. S. (2020). Traditional agricultural practices in India: An approach for environmental sustainability and food security. *Energy, Ecology and Environment*, *5*. https://doi.org/10.1007/s40974-020-00158-2

Prabhu, S., Chitale, V., & Jain, K. (2024). *Decoding India's changing monsoon patterns: A Tehsil-level assessment*. Centre for Energy Finance. https://www.ceew.in/sites/default/files/decoding-how-climate-change-is-changing-monsoon-rainfall-patterns-in-india.pdf

Pradhan Mantri Fasal Bima Yojana (PMFBY). (2019). *Crop insurance*. https://pmfby.gov.in/

Rezaei, E. E., Webber, H., Asseng, S., Boote, K., Durand, J. L., Ewert, F., Martre, P., & MacCarthy, D. S. (2023). Climate change impacts on crop yields. *Nature Reviews Earth & Environment*, *4*, 1–16. https://doi.org/10.1038/s43017-023-00491-0

Saha, P., Mahanta, R., & Goswami, B. N. (2023). Present and future of the South Asian summer monsoon's rainy season over Northeast India. *Npj Climate and Atmospheric Science*, *6*(1), 1–11. https://doi.org/10.1038/s41612-023-00485-1

Saleem, A., Anwar, S., Nawaz, T., Shah Fahad, Saud, S., Tanzeel Ur Rahman, M., & Nawaz, T. (2024). Securing a sustainable future: The climate change threat to agriculture, food security, and sustainable development goals. *Journal of Umm Al-Qura University for Applied Sciences*. https://doi.org/10.1007/s43994-024-00177-3

Statistical yearbook world food and agriculture 2023. (n.d.). https://openknowledge.fao.org/server/api/core/bitstreams/28cfd24e-81a9-4ebc-b2b5-4095fe5b1dab/content/cc8166en.html

Streimikis, J., & Baležentis, T. (2020). Agricultural sustainability assessment framework integrating sustainable development goals and interlinked priorities of environmental, climate and agriculture policies. *Sustainable Development*, *28*(6). https://doi.org/10.1002/sd.2118

Sunder, K. (2020, September 11). *The remarkable floating gardens of Bangladesh*. https://www.bbc.com/future/article/20200910-the-remarkable-floating-gardens-of-bangladesh

Suresh, A., Krishnan, P., Jha, G. K., & Reddy, A. A. (2022). Agricultural sustainability and its trends in India: A macro-level index-based empirical evaluation. *Sustainability*, *14*(5), 2540. https://doi.org/10.3390/su14052540

The Agrobiodiversity Index|International Treaty on Plant Genetic Resources for Food and Agriculture|Food and Agriculture Organization of the United Nations. (2024). https://www.fao.org/plant-treaty/tools/toolbox-for-sustainable-use/details/en/c/1312762/

The Weather Risk Management Facility (WRMF)|NDC Partnership. (2024). https://ndcpartnership.org/knowledge-portal/climate-funds-explorer/weather-risk-management-fa cility-wrmf#:~:text=The%20Weather%20Risk%20Management%20Facility%20%28WRMF%2 9%20is%20a

TOI News Desk. (2024, July 10). 557 Farmers died by suicide in Maharashtra in first 6 months of 2024: State government report. *The Times of India*. https://timesofindia.indiatimes.com/india/557-farmers-died-by-suicide-in-maharashtra-in-first-6-months-of-2024-state-government-report/articleshow/111637528.cms

UN Environment. (2018, June 2). *Andhra Pradesh to become India's first Zero Budget Natural Farming state*. https://www.unep.org/news-and-stories/press-release/andhra-pradesh-become-indi as-first-zero-budget-natural-farming-state

United Nations. (1992). *United Nations framework convention on climate change*. United Nations. https://unfccc.int/resource/docs/convkp/conveng.pdf

United Nations. (2021, October 21). *Madagascar: Severe drought could spur world's first climate change famine*. UN News. https://news.un.org/en/story/2021/10/1103712

World Bank. (2019). *Climate-smart agriculture investment plan for Ghana*. https://documents.worldbank.org/en/publication/documents-reports/documentdetail/3001615923 74973849/climate-smart-agriculture-investment-plan-for-ghana

World Bank. (2020, October 7). *Global action urgently needed to halt historic threats to poverty reduction*. World Bank. https://www.worldbank.org/en/news/feature/2020/10/07/global -action-urgently-needed-to-halt-historic-threats-to-poverty-reduction

World Food Programme. (2011). *The R4 rural resilience initiative*. https://www.wfp.org/r4-rural-resilience-initiative

Zachary. (2023, September 25). *System of rice intensification in India*. https://storymaps.arcgis.com/stories/08ec39b621e146f9b0cc2ec18fd06717

Index

Printed and bound by CPI Group (UK) Ltd, Croydon, CR0 4YY

28/09/2026

14971126-0002